living in the
anthropocene

LIVING IN THE
ANTHRO
EARTH IN THE

EDITED BY
W. JOHN KRESS AND JEFFREY K. STINE

FOREWORD BY
ELIZABETH KOLBERT

AFTERWORD BY
EDWARD O. WILSON

RICHARD B. ALLEY

SUBHANKAR BANERJEE

CARTER J. BRANDON

LONNIE G. BUNCH III

PAULA CABALLERO

KELLY CHANCE

ROBIN L. CHAZDON

LINDSAY L. CLARKSON

DOUGLAS J. MCCAULEY

SEAN M. MCMAHON

J. R. MCNEILL

KAREN E. MILBOURNE

ROB NIXON

ARI NOVY

RICK POTTS

STEVEN J. PYNE

POCENE

AGE OF HUMANS

G. WAYNE CLOUGH

WADE DAVIS

PETER DEL TREDICI

J. EMMETT DUFFY

FINIS DUNAWAY

JOHN GRABOWSKA

NAOKO ISHII

LUC JACQUET

IGOR KRUPNIK

THOMAS E. LOVEJOY

GEORGE E. LUBER

JOANNA MARSH

LISA RUTH RAND

PETER H. RAVEN

TORBEN C. RICK

HOLLY H. SHIMIZU

CORINE WEGENER

SCOTT L. WING

In association with
Smithsonian Institution Scholarly Press

Smithsonian Books
WASHINGTON, DC

This book may be purchased for educational, business, or sales promotional use. For information, please write: Special Markets Department, Smithsonian Books, P.O. Box 37012, MRC 513, Washington, DC 20013

Published by Smithsonian Books
Director: Carolyn Gleason
Managing Editor: Christina Wiginton
Project Editor: Laura Harger
Edited by Juliana Froggatt
Designed by Jody Billert
Typeset and indexed by Scribe, Inc.

Library of Congress Cataloging-in-Publication Data

Names: Kress, W. John, editor. | Stine, Jeffrey K., editor.
Title: Living in the anthropocene : earth in the age of humans / edited by W. John Kress and Jeffrey K. Stine ; foreword by Elizabeth Kolbert ; afterword by Edward O. Wilson ; essays by Richard B. Alley [and others].
Description: Washington, DC : Smithsonian Books, in association with Smithsonian Institution Scholarly Press, [2017] | Includes bibliographical references and index.
Identifiers: LCCN 2016059274 | ISBN 9781588346018 (hardcover : alk. paper) | ISBN 9781588346025 (ebook)
Subjects: LCSH: Human ecology. | Nature—Effect of human beings on. | Global environmental change.
Classification: LCC GF75 .L575 2017 | DDC 304.2—dc23
LC record available at https://lccn.loc.gov/2016059274

Manufactured in the United States of America
21 20 19 18 17   5 4 3 2 1

CONTENTS

III | Responding to Change

IV | Visual Culture

V | The Way Forward

FOREWORD

ELIZABETH KOLBERT

Credit for coining the word *Anthropocene* is usually given to the Dutch chemist Paul Crutzen, whose other accomplishments include, quite literally, saving the world. Back in the 1970s, Crutzen was one of the first scientists to recognize the dangers of ozone-depleting chemicals; for his work in this area, he shared a Nobel Prize in 1995. A few years later, he was attending a meeting at which the chair of the session kept referring to the Holocene, the geologic epoch that began with the end of the last ice age, roughly twelve thousand years ago. It occurred to Crutzen that the term no longer made sense.

"Let's stop it," he remembers blurting out. "We are no longer in the Holocene; we are in the Anthropocene." The immediate reaction to this comment was a sort of stunned silence. Afterward, when the group took a coffee break, someone suggested that Crutzen patent the term.

Whether or not geologists officially recognize *Anthropocene*—they are still debating whether textbooks should be altered—the word clearly captures something essential about our time. We live in a world dominated by humans. Possibly people already began changing the atmosphere thousands of years ago, with the invention of agriculture. Quite certainly, we started to do so once we figured out how to burn coal and oil. We are, in effect, now running geologic history backward, taking carbon that was buried underground over the course of tens of millions of years and pouring it back into the atmosphere in a matter of decades. As a result, we are rapidly changing the climate and altering the chemistry of the oceans. At the same time, we are draining freshwater aquifers, mowing down forests to plant monocultures, altering the global nitrogen cycle, and driving other creatures extinct at rates hundreds or thousands of times higher than the geologic norm. The legacy of the Anthropocene will be, in human terms at least, permanent. Once you lose a species, you do not get it back.

The more we understand about our impacts, and our impacts' impacts, the more urgent the questions become. What should we do with this knowledge? Should we scale back our influence? Can we? What do those alive today owe to future generations? What about to the millions of other species with which we share the planet?

The essays that follow take up these great questions. They are written from a wide range of perspectives—scientific, social, artistic, and economic—by men and women who have thought deeply about living in the Anthropocene. Although people created this new age, it does not follow that we control it. We find ourselves in the unfortunate situation of being more powerful than we ought to be and, at the same time, not as powerful as we might wish. What we are belatedly realizing is that turning away from the problem—or, really, problems—is not an option. The best hope we have is to acknowledge the enormity of the challenge and try to fashion responses that are commensurate in scale.

Crutzen once told me that the act of naming the Anthropocene was intended to serve as an alarm. "What I hope," he said, "is that the term *Anthropocene* will be a warning to the world." As this volume demonstrates, in this he succeeded.

INTRODUCTION

W. JOHN KRESS AND JEFFREY K. STINE

At a rate unprecedented in the recent past, our planet has been experiencing a multitude of dramatic and far-reaching changes—in the vegetation covering the land, in the chemistry of the oceans, in the concentration of atmospheric greenhouse gases, and in global temperatures. Most disturbing is that these transformations, with their profound effects on plants, animals, and natural habitats, are primarily the result of human activities. Earth, *our* Earth, has, of course, always been a planet in flux, with the scope and pace of its changes varying dramatically over its 4.55-billion-year history. In the narrative of this deep geologic timescale, the evolutionary forebears of humans are very recent arrivals, appearing roughly six million years ago, while the history of our own species, *Homo sapiens*, extends back a mere two hundred thousand years. Fortuitously, as the last ice age came to an end some twelve thousand years ago, Earth entered a period (known today as the Holocene) characterized by a relatively warm and stable climate and with conditions well suited to humans. Profiting from such hospitable circumstances, our ancestors developed agriculture, which served as a precursor to large settlements, eventually to civilizations, and then to the Industrial Revolution, which began around the year 1800.

Modern societies—and the technological systems that sustain them—thus arose entirely within the benign envelope of the Holocene, encouraging a false assumption that such conditions would endure, unaffected by the presence and actions of humans. By the late twentieth century, however,

1

that assumption had started to fray. Signs of serious, large, and rapid global environmental change were increasingly evident—from climate disruption and ocean acidification to deforestation and biodiversity loss—and the rate and scale of those alterations have attained levels unseen since the origin of humans. Agricultural, industrial, and other human activities have conjoined to modify atmospheric, geologic, hydrologic, biospheric, and other Earth systems to such an extent that scientists have proposed our time as the beginning of a new geologic epoch: the Anthropocene, or Age of Humans. From altering the migrations of plants, animals, and people to exacerbating the rise and spread of infectious diseases, the environmental impacts of human activity have never been greater. And continuing growth in population, urbanization, and societal conflict will intensify those impacts, reinforcing the fact that nature can no longer be viewed in isolation from the human world.

Thanks to research in many fields of study, the contours of these planetary impacts are becoming clearer, making it incumbent upon all of us to consider the implications of what we know (and do not know) about the trajectory of Earth's future.

Living in the Anthropocene presents thirty-two original essays by a roster of distinguished authors from a wide range of disciplines. By blending the diverse perspectives and expertise of environmental scientists, historians, archaeologists, anthropologists, economists, art historians, and documentarians, this book seeks to advance understanding of the complex causes and consequences of human-induced environmental change. After describing the current state of our planet, *Living in the Anthropocene* proceeds to address the drivers of that change, the adaptations of both humans and nature to that change, and the depictions of that change by visual artists. The book concludes with critical perspectives on paths forward as the rate of global change increases.

The authors converge upon several themes, such as the necessity of adopting a deep-time perspective to grasp the geologic significance of the Anthropocene. The use of geologic and evolutionary timescales helps to clarify both how human-driven impacts, such as refashioning the planet's carbon and nitrogen cycles, parallel past alterations that marked major transitions in Earth history and how the biogeochemical effects of anthropogenic changes will persist for thousands, if not tens of thousands, of years. Paleontological knowledge of Earth's five previous mass extinctions raises awareness of just how difficult it is for life to survive rapid planetary change, which argues for the importance of curbing the rate at which human activities are diminishing global life-support systems. This broader temporal perspective also highlights how earlier periods of high environmental variability shaped human evolution, honing our keen ability to adapt.

Many of these essays explore the Anthropocene's cultural, social, political, and economic dimensions. Drawing from a wide range of past and present examples, they explain that the whole human species is not responsible for the negative effects of global environmental change, just as not all groups are equally threatened by the ramifications of those changes. Indeed, it is a cruel irony that social groups that have contributed the least to such planetary effects are often those most seriously harmed by the consequences. In today's world, the links among economic inequality, social injustice, and environmental degradation are unmistakable. Indigenous peoples, disproportionately jeopardized by global environmental change, also hold deepseated knowledge of their homelands and of expressions of the early effects of climate change. Their plight presents the world with a moral challenge, while the manners in which they have adapted offer examples of viable options for the future.

Throughout human history, visual representation has helped to record environmental change as well as spark imagination. Getting people to recognize and understand the defining characteristics of the Anthropocene often involves translating science to general audiences in countless settings around the world. Both visual culture and history can play important roles here. The impact of art, including the emotions it invokes, will be decisive in how our societies will exist in the midst of global change.

All of the authors here take a frank look at humanity's future, refusing to downplay the difficulties. Their use of the term *Anthropocene* is not a proxy for saying that the world is facing an environmental crisis. Instead it is meant to suggest something far more profound and lasting; *crisis*, after all, denotes a temporary situation potentially remediable through sacrifices and coordinated efforts. Earth has already entered a new epoch, one in which the effects of many human-induced alterations will continue for generations. The current ten million species, including our own, are the descendants of the billions of species that have existed in the past, and new species will eventually evolve to take our place. Until then, we must learn how to live in the Anthropocene.

A CHANGING

PLANET

A s the name *Anthropocene* suggests, the impact of human activities has reached global proportions. The physical and biological transformations now taking place may be equivalent in magnitude to the major environmental transitions that marked significant geologic turning points in the distant past, such as the start of the Eocene, the Paleocene, and the Holocene. The bio-geochemical consequences of anthropogenic change are both far reaching and profound, as are its social, cultural, political, and economic effects. Gaining a full understanding of the complex causes and implications of these planetary alterations therefore requires a blending of perspectives from many fields of study. The start of the Anthropocene remains a matter of debate, although the mid-twentieth century—which witnessed a phenomenal escalation of a wide spectrum of environmental indicators (the so-called Great Acceleration)—is widely acknowledged as a significant turning point in human-induced environmental change.

The most obvious transformations have taken place within terrestrial habitats, but the oceans have also been intensely affected by acidification, warming, mass extinctions of marine mammals, depletions of wild fish, industrialization of undersea landscapes, and intensification of plastic pollution. Natural environments, whether terrestrial or marine, may never be restored to anything resembling their prehuman conditions. Our evolutionary success as the most recent bipedal hominid species has been fostered by our ancestors' exceptional ability to adapt to changing environments, often by altering our surroundings through the use of technology and social organization. Since the Industrial Revolution, economic growth has become an end in itself for much of the world, and the ramifications of this worldview can be seen in the environmental challenges characteristic of the Anthropocene.

❧ THE ADVENT OF THE ANTHROPOCENE

J. R. McNEILL

In 1944, the Hungarian-born social scientist Karl Polanyi published a tur-
gid, difficult, and detailed book entitled *The Great Transformation*. In it he
showed that markets, as the chief means of distributing goods and services
among populations, had not dominated any societies before 1700. They were
not in any sense natural or the reflection of innate human desires. Alternative
arrangements had flourished in the past, even if, by 1944, it was hard for
most people to conceive of a functional economy not based overwhelmingly
on markets. Polanyi showed that the prevalence of markets was comparatively
recent and had required certain changes in societies and politics to triumph.

An homage to Polanyi, the term *Great Acceleration* refers to the sharp
mid-twentieth-century uptick in the rate of ecological change around the
globe. To most of us, it now seems normal, almost natural, that humans
should exert vast influence over the biosphere and basic biogeochemical
systems. That is because we cannot remember a world in which this was not
true. But in fact that condition is novel and a bizarre departure from the
arrangements that governed the human place in the biosphere for the first
two hundred thousand years of *Homo sapiens*'s career. The concepts of the
Great Acceleration and the Anthropocene can help put into high relief
the uniqueness of modern times and remind us that what we easily misun-
derstand as normal and natural is indeed anything but.

A few examples will underscore the reality of the Great Acceleration.
Our forebears took many millennia to reach a population of one billion,
which occurred around 1800 or 1820. It then took more than a century for
human numbers to reach two billion (around 1930). Soon, in the middle
of the twentieth century, an unprecedented frenzy of survival and reproduc-
tion began, and the human population tripled in the span of one lifetime.
It reached three billion by 1960 and four billion by 1975, and thereafter it
added a new billion every twelve to thirteen years. Meanwhile, total energy
use quintupled between 1950 and 2015. In those sixty-five years, people

burned a quantity of fossil fuels that had taken 150 million years to accumulate. More than three-quarters of the anthropogenic greenhouse gas emissions in human history occurred during those sixty-five years. The world's motor vehicle fleet grew like kudzu, from forty million to nine hundred million, over the same time span. These and many more accelerating trends made the post-1950 world very different from all that had come before it. Collectively, they vaulted us into a new time period in both human history and the history of Earth: the Anthropocene.

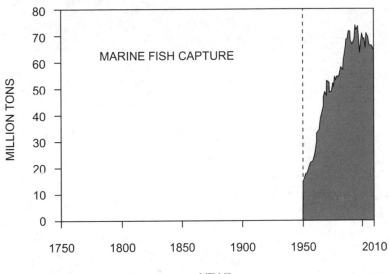

Figures 1–8. These eight graphs collectively show that the middle of the twentieth century was a transitional moment for several indicators, or driving forces, of global environmental change. For those who embrace the concept of the Anthropocene and endorse a mid-twentieth-century birthday for it, these graphs provide persuasive evidence. Several more such charts appear on the website of the International Geosphere-Biosphere Programme. (OECD: Organisation for Economic Co-operation and Development; BRICS: Brazil, Russia, India, China, and South Africa.) From W. Steffen et al., "The Trajectory of the Anthropocene: The Great Acceleration," *Anthropocene Review* 2 (2015): 81–98. Used with permission of Sage Publishers.

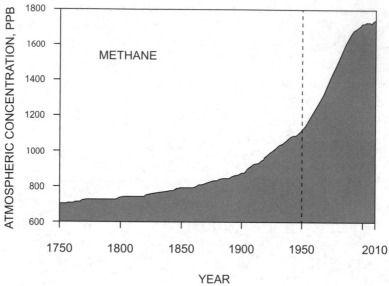

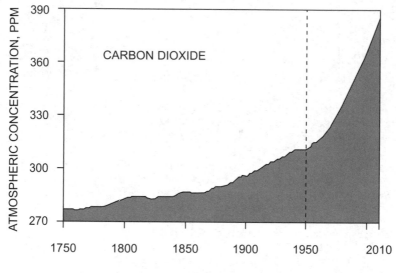

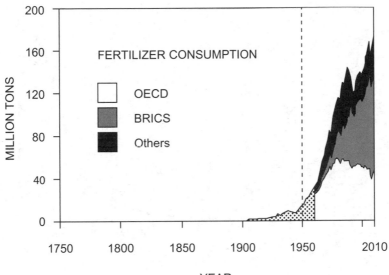

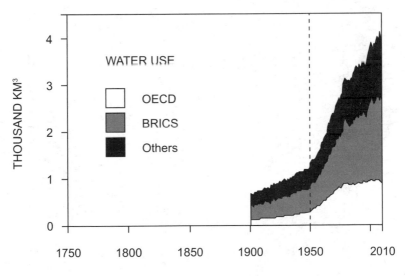

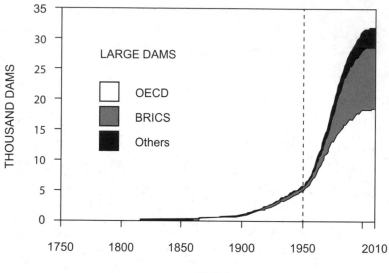

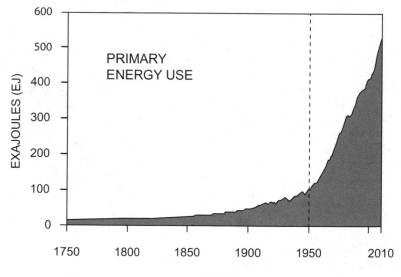

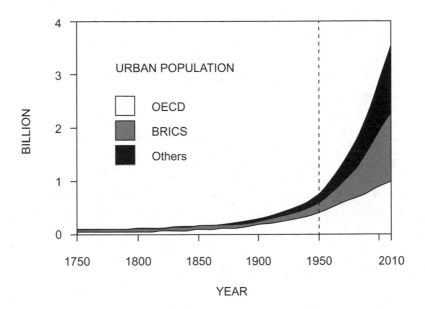

URBAN POPULATION

☐ OECD

▨ BRICS

■ Others

"The Anthropocene" at present means all things to all people. Arguments simmer about how old it is. The weight of the evidence suggests that the Anthropocene began in the mid-twentieth century, but some scientists argue for a much older Anthropocene, beginning with the harnessing of fire, the late Pleistocene extinctions, or the dawn of agriculture. Others prefer an Anthropocene dating to the Columbian Exchange (beginning in 1492), when sailors carried ark-loads of species of animals, plants, and microbes from one continent to another, leaving a lasting mark in the paleontological record. Still others see the Industrial Revolution (ca. 1780–1850) as the decisive period that sets the Anthropocene apart, thanks to the rapid adoption of fossil fuels and the subsequent greenhouse gas emissions that were generated on an ever-larger scale in those years.

Arguments about the antiquity (or novelty) of the Anthropocene are also arguments about its essence. For most of those who regard atmospheric chemistry as the most relevant marker of global change, an Anthropocene that begins with industrialization makes the most sense. For those who require that the Anthropocene leave a signal in the paleontological record and who see the history of life on Earth as the key variable in demarcating intervals of time, the late Pleistocene extinctions or the Columbian Exchange make more sense.

If one tries to mesh all the relevant variables and to avoid privileging any particular discipline's outlook and preferred forms of evidence over others, the Great Acceleration seems to be the best birth date for the Anthropocene. Taking a "basket of variables" approach, as the International Geosphere-Biosphere Programme recently did, underlines the many respects in which the mid-twentieth century marks a point of inflection on rising curves. Beyond the upticks in rates of population growth and energy use, which together may lie at the heart of the matter, a long list of relevant variables describe a similar trajectory: carbon emissions, methane emissions, fertilizer consumption, ocean acidification, and nitrogen loading of coastal waters, among others. It is the collective weight of all these variables that separates the Anthropocene from what came before, rather than any single one. Their simultaneous and interwoven acceleration trajectories, beginning about 1950, blasted us into the Anthropocene.

The Great Acceleration is doomed. The remarkable trends that most of us have lived with all our lives will not last long. In some cases, finite supply is the issue. The world does not have enough fresh water to allow another quadrupling of water withdrawals (as occurred from 1950 to 2010). There are not enough good sites left to support another sextupling of the world's large dams. There are not enough fish to permit another quintupling of the marine fish catch. In other cases, saturation or equilibration is the issue. The proportion of the global population living in cities, now at more than 50 percent, cannot more than double (as happened from 1950 to 2010). Total human population could in theory triple once more, as it did in that same period, but no one imagines it will, because urbanization, formal female education, and other social changes have sharply reduced couples' reproductive ambitions. Globally, human fertility is only a little more than half of what it was in 1970. So for all these reasons and several others, the Great Acceleration will come to a close. Indeed, many of its trends have already leveled off (dam building, marine fish catch) or begun to decline.

The Anthropocene, however, will live on. Even if the human population starts to fall some fifty or sixty years hence (as some speculate it will), even if by 2075 we have banished fossil fuels to the margins of a low-carbon energy system, even if green parties win every election, the Anthropocene will live on. That is because, for a long time to come, there will still be billions of people using the global environment as a source of materials and a sink for wastes, even if at restrained rates, and for a long time to come, the carbon already emitted into the atmosphere will continue to trap heat and warm Earth's surface and its oceans. Less certainly, increasing skill in manipulating human and other genomes could give a new tint to the Anthropocene, raising the efficiency with which we eliminate some species (pesky mosquitoes, perhaps) and alter others.

Thus, of all the possible understandings of the Anthropocene, the one that best matches the evidence is the post-1950 Anthropocene, born of the Great Acceleration. Historians may leave it to others to reflect upon whether the Anthropocene is on balance a good or a bad thing and how long it might last. Those questions require a clear look into the future, and historians have trouble enough seeing into the past.

THINKING LIKE A MOUNTAIN IN THE ANTHROPOCENE

SCOTT L. WING

Aldo Leopold's posthumously published book *A Sand County Almanac* (1949) includes an essay entitled "Thinking Like a Mountain." In it, he recounts shooting a wolf as a young man. Watching the "fierce green fire" in her eyes die helps him to consider the deeper meaning of the wolf's existence—or, as he puts it, to think like a mountain. In this essay of only 878 words, Leopold gave twentieth-century conservationists two powerful but distinct metaphors. The fierce green fire speaks of the emotional effect of losing the wildness that wolves represent. Thinking like a mountain suggests a more detached point of view but one that appreciates the interplay of the wolf and its ecosystem. In the decades after he shot the wolf, Leopold writes that he has "watched the face of many a newly wolfless mountain, and seen the south-facing slopes wrinkle with a maze of new deer trails . . . seen every edible bush and seedling browsed, first to anaemic desuetude, and then to death." What he has learned, and what the mountain has always known, is that extirpating predators leads to explosive growth of herbivore populations, overbrowsing, reduced vegetation, and loss of soil—in short, a cascade of unintended, long-term consequences considered undesirable by the humans who initiated them. Leopold's essay stretches the minds of readers unaccustomed to thinking about the biological and landscape changes that can be precipitated by removing a keystone predator from an ecosystem.

The thinking mountain of the essay's title recognizes ecological complexity, but its spatial frame is regional rather than global and its time scale more human than geologic. Looking back, though, we see that *A Sand County Almanac* was published near the onset of the dramatic upward inflection in human resource use and global effects often referred to as the Great Acceleration, making it one of the first conservationist texts of the Anthropocene. Three score and eight years is not a long time to watch a new epoch unfold, but the changes in the environment that have taken place in those years may justify a reevaluation of Leopold's classic metaphor.

The idea behind the Anthropocene is conceptually simple: the changes we are now making to Earth are on a par with the shifts in the global environment that mark the beginnings of previous geologic time periods. (By convention, stratigraphers define each period of the geologic time scale by its beginning, or base, with the end of one being defined by the beginning of the succeeding period. No gaps or interregna are allowed in geochronology.) Although there is plenty of scholarly debate about naming a formal Anthropocene epoch, there is no doubt that the changes humans have been causing to Earth systems since the mid-twentieth century are comparable in type and magnitude to past changes that we use to recognize the beginnings of new phases of Earth history. Indeed, some of these earlier changes reflect perturbations of the very same processes we are now altering, even though their rates and ultimate causes are different.

The beginning of the Eocene epoch (approximately fifty-six million years ago), for example, was accompanied by a major release of carbon, with attendant global warming and ocean acidification, an event called the Paleocene-Eocene Thermal Maximum, or PETM. Although the amount of carbon released at the onset of the PETM was greater than the amount that humans are likely to generate except in the most extreme future scenarios, the rate of carbon release today is probably ten times faster. The rapid shift in carbon chemistry (the ratio of ^{12}C to ^{13}C) at the onset of the PETM now forms a convenient marker for the base of the Eocene, as do the changes in the composition of fossil faunas and floras that it caused. The effects of the PETM on Earth's climate and biota lasted for more than a hundred thousand years. In a parallel fashion, the change in the ratio of carbon isotopes caused by burning fossil fuels will form a permanent sedimentary marker for recognizing an Anthropocene epoch, as will the massive changes in biodiversity, sedimentation, nutrient supply, and ocean chemistry that are resulting from human activities.

Since 1949, we have gained much greater insight into how the integrated Earth-life system works. Geologists have contributed to this understanding by reconstructing past global environmental changes and the interaction of environmental change with life. Earth system scientists have developed the ability to monitor environmental change globally in real time—giving us a sense of how, and how fast, the Earth's system is shifting. Powerful computers can now simulate the interconnected processes that influence the global environment, producing predictions of future change. As a result, we know not only that we are increasing carbon dioxide (CO_2) in the atmosphere every year but also that much of that carbon dioxide will still be in the atmosphere in the year 3000, that most of the resulting increase in temperature and sea level will still be in force in the year 12,000, and that a full return to background carbon dioxide won't occur for more than a hundred thousand years. Human effects on the global environment are large and unprecedented and are producing a welter of unintended consequences. Even more important, though less widely appreciated, is that these human-induced changes will persist for far longer spans of time than the historical, or even archaeological, record that shapes our thinking.

The revolution in Earth system science that has taken place since 1949 should have radically altered our sense of who we are. The new insights haven't undone the Copernican revolution, which removed us from the center of the solar system, or the Darwinian revolution, which revealed that we exist because of the same evolutionary processes that generated all life on Earth, but current understanding of the Earth's system does show that we are no longer a bit player in the story of this planet, and that the influence of our actions now will change the global environment for at least hundreds of human generations to come.

The distant future will never be as salient as tomorrow. Nevertheless, in the Anthropocene we have to learn to anticipate and adapt to the long-term consequences of our actions and to avoid taking actions with dire consequences. Although some imagine that the rapid changes we are causing in the global environment are an existential threat—that we will drive ourselves extinct—this scenario is a very remote possibility. A planetary population that has more than doubled since the mid-twentieth century and will likely reach more than nine billion in the next thirty years is the opposite of endangered. Visions of human extinction are more escapism than reality: believing in the apocalypse means you don't have to plan for the future. The apocalypse may be a dark fantasy, but even without an existential threat there is enormous potential for human misery in a hotter future with flooded coastlines, longer droughts, more violent storms, fewer forests and coral reefs, and fewer species. Only the most resigned cynic or bloated egotist could conclude that short-term benefits to current generations trump the long-term costs to the many generations to come.

If it is clear that we will not "destroy the planet," as it is sometimes put, it is also clear that we don't have the capacity to return it to a "state of nature," if that implies no human influence. The seven billion of us now on the planet use between a quarter and a half of global net photosynthetic productivity and have left far less than half of Earth's land area in a wild or seminatural state. Even with dramatically more sustainable practices, the portion of the planet and its resources devoted to supporting humans will almost certainly increase as the human population continues to swell. Any attempt to rapidly return the planet to an imagined prehuman state would cause enormous human suffering. Furthermore, change is inevitable. Earth history shows us that climate veers from warm to cold, that the composition of the atmosphere changes, and that cataclysms happen. Rather than justifying anthropogenic global change, however, this perspective shows just how hard it is to survive such rapid change. Slowing the rate at which we alter global life-support systems may seem a modest goal, but it would allow more time for adaptation to changes that cannot be avoided, as well as time for natural negative feedbacks to counteract some of the changes we are forcing.

Researchers of the Earth's system have been focused, appropriately, on developing a better understanding of the vast and interconnected processes that create our environment, and they have made a great deal of progress since the publication of *A Sand County Almanac*. Although there are many problems left to solve, knowledge about planetary life-support systems has progressed far more rapidly than society's willingness to use this knowledge. The biggest challenge facing humanity is that our political, social, and economic systems are shortsighted. Long-term planning typically considers years or decades, but the global environmental processes we are now influencing play out over centuries, millennia, or more. We need to instill a sense of geologic time into our culture and our planning, to incorporate truly long-term thinking into social and political decision making. This is what "thinking like a mountain" should come to mean in the Anthropocene. If we succeed in transforming our culture, residents of the later Anthropocene will look back on the early twenty-first century as a time of human enlightenment, when people learned to truly think like mountains by anticipating their long-lasting and complex effects on the world. The moral price of our knowledge and influence is that we must use them responsibly to shape our future and that of the planet. We must take a much longer view of our legacy than we ever have before.

⚘ THE UNDERWATER ANTHROPOCENE
DOUGLAS J. McCAULEY

S cientific dialogues on the Anthropocene rarely extend below the high-tide line. This terrestrial bias is perhaps justifiable, as we have been altering terrestrial ecosystems since the African diaspora gained momentum about fifty thousand years ago. Today, croplands and pastures take up about 40 percent of Earth's land surface, while the forty million miles of road (a distance equivalent to 165 trips to the Moon) that we have laid out across the world have left less than 10 percent of the planet's land surface remote. The terrestrial portion of the world has been brought unambiguously under the dominion of our species.

By almost all measures, however, the mark of the Anthropocene has been lighter in the oceans. California, my home, provides an illustrative example. Humans assisted with the extirpation of terrestrial megafauna (e.g., eleven-ton mammoths, ground sloths more than ten feet tall) from the region about fourteen thousand years ago. We then proceeded to drive California's wolves and grizzly bears extinct (the latter our state animal and flag symbol). But today, just offshore and within eyesight of metropolitan skylines, thirty-three-ton gray whales undertake one of the longest mammal migrations on the planet, 550-pound giant sea bass vocalize at divers, and white sharks investigate the palatability of about one and a half California beachgoers annually. While deeply altered, the oceans retain a wildness that has become rare in much of the terrestrial world.

What delayed and muted the arrival of the Anthropocene in the oceans? The simple answer is that it is harder to change the oceans—at least for us terrestrial apes. Examples of nonhuman great apes affecting aquatic ecosystems are uniformly underwhelming: orangutans can catch disabled catfish, and bonobos scoop up the occasional aquatic animal when swamp foraging. But because we humans rely more on our brains than on tooth or claw, we eventually overcame the significant physical barriers that normally prevent terrestrial animals from hunting efficiently in ocean ecosystems. We invented our first deep-sea fishing technologies (e.g., bone fishhooks) and were catching pelagic fish at least forty thousand years ago. But it wasn't until after the Second World War, when we repurposed wartime marine technologies to industrialize fishing fleets, that we profoundly amplified our impact on the oceans and arguably first wet the feet of our global human footprint.

The International Union for Conservation of Nature (IUCN) recognizes more than six hundred species extinctions on land in the past 515 years, but only fifteen in the oceans. While this pattern genuinely reflects the late start of the marine Anthropocene, measurements of ocean extinction must be viewed as minimum estimates. Just as it is harder to cause extinction in the oceans, it is also much harder to detect marine extinctions. It took us seventy-three years to find the *Titanic* after it sank—and she weighed fifty thousand tons and was perhaps the most famous ship in all of history. It is easy to imagine that a cryptic marine species, such as a flatfish or goby, could go extinct without notice.

Measures of outright global extinction, by themselves, are insufficient barometers of anthropogenic change. Many extant marine species have been massively depleted in number both purposefully (e.g., highly priced and prized bluefin tuna) and accidentally (e.g., sea turtles as bycatch). Precipitous declines in the abundance of terrestrial species such as amphibians, bees, and bats are widely known, but drops of equal or greater intensity have recently been described for marine fauna: seabird species have declined by about 70 percent, numerous shark species by more than 90 percent, and certain great whale species by 80 to 90 percent.

One proposed start date for the terrestrial Anthropocene is about eleven thousand years ago, when key human populations switched from hunting and gathering to farming. A similar game-changing transition occurred in the oceans in 2014, when it was estimated that, for the first time, humans consumed more fish that came from aquaculture than from the wild. Throughout history, the oceans have served as publicly accessible seafood sections full of free-range meat, but the potential for wild terrestrial ecosystems to regularly provision humanity in this fashion went extinct in most parts of the planet hundreds of years ago. A repetition of this history in marine eco-systems would represent a radical shift in our relationship with the oceans.

Another transformative change in the Anthropocene oceans is the emergence of the Marine Industrial Revolution: a shift, now under way, from focusing on the capture of marine wildlife for consumption to using marine resources and marine real estate to foster new marine industries. The Marine Industrial Revolution is well exemplified by the explosive growth of marine mining, marine power generation, desalination projects, aquaculture, oil and gas extraction, and coastal construction. While much of this new ocean industry positively stimulates economic growth and helps meet food and energy shortfalls, it also ups the ante on how humans change the oceans. We have graduated from harvesting marine species to harvesting marine habitats.

The Anthropocene palpably manifests itself as colorful flecks in the cod end of plankton nets and in grabs of deep-sea sediment. Plastic pollution has become a near ubiquitous constituent of our modern oceans. We take about five million tons of tuna from the global oceans annually—and put back two to three times that amount of plastic. This plastic is making its way into marine food webs (for instance, it is estimated that 99 percent of seabirds will be swallowing plastic by 2050) and even onto our own dinner plates (25 percent of fish in market surveys contained plastic or fiber debris).

Like all parts of Earth, from rocks to human tendons, the tissues of animal life in the oceans (e.g., shark vertebrae, coral skeletons) were chemically marked by aboveground nuclear weapons testing during the 1950s. Bomb carbon, however, remains only one of a diverse array of indelible signatures of the Anthropocene left in our oceans. Increased industrial activity, for example, has fueled dramatic and potentially deleterious increases in the mercury levels of top marine predators, including albatross, whales, and seafood-eating humans. The 2011 Fukushima nuclear accident, too, marked a vast section of the Pacific with its eastward-dispersing chemical fingerprint. The raw power of humanity to write our history into the very bodies of marine life and the essence of the waves is impressive—and deeply disconcerting.

Nothing happens fast in a 352-quintillion-gallon water bath—and still we have begun to alter basic physical elements of the global ocean. Human-caused climate change is having well-known effects on ocean temperature, acidity, and sea level state, but it is also predicted to exacerbate ocean deoxygenation, perturb coastal upwelling, and alter patterns of ocean circulation. A steadily rising Anthropocene ocean that is hotter, harder to breathe in, and more acidic presents obvious challenges to the future of marine life. The Anthropocene has definitively begun to wash from the land into the oceans, and although its arrival has been delayed and its effects are still less intense there, humanity has already fundamentally altered the ecology, chemistry, and physics of the oceans.

As the first impacts of the marine Anthropocene come into view, so too do the first consequences of living with an altered ocean. Climate-induced shifts in oceanography and weak governance will disproportionately degrade fisheries in poor tropical regions where access to highly nutritious marine foods is just barely keeping myriad malnutrition diseases at bay. Loss of marine wildlife has been linked to increases in insidious social injustices, such as human trafficking and piracy. Degradation of ecosystems also imperils the sustained provisioning of the $2.5 trillion in goods and services that come to us yearly from the oceans.

Is there reason to be optimistic about our potential to constructively engage the arrival of the Anthropocene in the oceans? Definitively yes. Emerging marine industry can be intelligently managed to provide clean energy and new resources without deleteriously usurping ocean ecosystems. Prudent management of wild fisheries can ensure that we can have our marine biodiversity and eat it, too. If we meaningfully follow through on recent groundbreaking global promises to slow climate change, we can buy ocean animals time to adapt to a changing ocean.

It is precisely because the Anthropocene has only just begun in the oceans that we retain a hopeful, meaningful, and valuable opportunity to control how it evolves. The inextricable links between human health and ocean health dictate that much will be determined by how we decide the Anthropocene will unfold in the sea.

❦ WHAT WILL IT MEAN TO BE HUMAN?

RICK POTTS

The narrative of human origins is the heritage of all people alive today. It is a six-million-year journey of evolutionary forebears, all of whom walked upright and possessed significant composites of the features that define our version of humanity, *Homo sapiens*. Of some two dozen lineages of bipeds currently known to science, we are the last one standing. All the prior ways of being human have become extinct. Our species' evolved capabilities set the foundation for how we have, somehow, survived and ultimately spawned a radically altered world. The question of our collective future is this: What will it mean to be human?

For untold millennia, human survival has depended on altering things: making a tool, building a fire, constructing a shelter. More than 2.5 million years ago, our tool-bearing forebears carefully struck one stone on another and inaugurated a way of life dedicated to modifying the world within reach. Even such primeval manipulations required social responsibilities as these predecessors carried food to others, initiating that beautiful oddity of human anticipation called sharing. The discovery of a fossil skull of a decrepit human ancestor from 1.8 million years ago, who survived toothless for many years, has further shown that early toolmakers, at least on occasion, extended care to the incapacitated.

For hundreds of thousands of years, the predecessors of *Homo sapiens* made hand axes and other reliable implements for simple purposes. By three hundred thousand years ago, matters had begun to change. From African excavations come the oldest hints of Stone Age innovation: smaller implements, a diverse tool kit, and a lethal invention—the first stone points that could be launched through the air to deeply penetrate their targets.

The reshaping of humanity, however, extended beyond technological know-how, as a growing sentience of others also emerged at this time. Social networks and symbols began to knit together these new-styled toolmakers. Highly valued obsidian stone was shared over long distances among an ever-expanding web of neighbors. These ancestors also collected pieces of red hematite and black manganese. Red and black thus became significant colors, denoting symbolic motives and practices around which cultures began to diversify. The human venture as expressed by peoples around the globe today had its start in that early era.

How did such fundamental transitions in the human way of life occur? Environmental dynamics seem to have played a role. The vital benchmarks in human evolution coincided with times of unstable climate and tectonic shifts of the landscape. The tales told by African fossil remains and prehistoric artifacts imply that the origins of technological innovation, symbolic thought, social networks, and our species all were impacted by the vicissitude and novelty of a restless environment.

Development of the social and mental foundations of human culture sped up the adjustment to new conditions. The most exaggerated expressions of human sociality today—think of massive gatherings in rituals of celebration, competition, and mourning—are possible because such vast numbers of people can plan by imagining an imminent future they have not yet witnessed. We hold such abstractions in our collective brains and can act accordingly. Values, a sense of life's purpose, combat, and moral acts are all conveyed within the symbolic universes we create and are made possible by an ability to conceptualize and produce codes about prospects neither visible nor immediately present. Perhaps nothing is more distinctive of our species than our recognition of forces beyond direct human senses—gleaned by curiosity and imagination—that are then made real and crucial motivators of life and meaning.

These central aspects of human life reflect psychological and societal responses to uncertainty, attempts to make sense of a world prone to change. The ability to create meaning—whether religious, ethical, artistic, philosophical, or scientific—is part of the tool kit that enabled *Homo sapiens* to meet disruption in the external and social domains. The more we investigate the survival challenges of human origins, the more we see that the elaboration of human sociality, technology, and influence came about in an unnurturing world, inculcating in us the dual urges to explain and to secure gain.

Now we have fashioned a new era of uncertainty by adding our bold signature to this dynamic Earth. Surely statistics and other numbers matter as measures of our intellectual grasp of the likelihoods ahead. Yet to understand humanity also requires our wisest comprehension of the mental and social proclivities of our species, a recognition of both our shared origin and our cultural differences, appreciating our capacity for problem solving and the range of emotions aroused by dearth and plenty.

Dwindling reservoirs and waterways, rising seas that threaten homes, droughts that prompt hunger: such possibilities evoke lessons from the past about human well-being. When essential needs are unmet, nothing less than unrest, fear, and degradation of the better angels of our nature soon follow. We dream, as all predators do, and in those dreams lie utopias and nightmares. Particularly haunting is the loss of confidence in our neighbor, who is now everyone on Earth, since our social web has now become global. In times of change and uncertainty, the greatest potential casualty, therefore, is the attention paid by neighbors in times of need. A rising tide of mistrust would mean that the universal of self-interest has begun to defeat the universal of sharing and caring. Both are part of human nature, and while there is an intellectual tradition that associates human nature with raw biological instincts in battle with civilized humanity, our basest impulses are actually born of learning and culture that recruit reason and conscience to the cause of malice.

So where is there hope? And what *will* it mean to be human? These two questions are deeply entwined.

It has been said that civilization, especially the tendency to extend tolerance to larger numbers of individuals, is a compensatory consequence of the will to harm and kill one another. While such a casual hypothesis may do more to provoke than to enlighten, it suggests that reactions to an opposing impulse can arouse profound change within us: benevolence may arise from the dust of cruelty, knowledge as a reaction to ignorance, and perhaps even wisdom when confronted with lunacy.

Altered climate, eroded landscape, misused water, and wasted food plainly will impact people unevenly. Injury will be unequal relative to wealth, crowding, security, and geography. A sense of injustice and resentment is inevitable, and even modest projections suggest that hundreds of millions of people will need to relocate and recast the status quo of their lives. In this new era, our reactions to strife and the values we adopt will define our humanity.

The question of our future depends on whether cooperation and altruism will counter self-interest and apathy, and whether empathy and connection can overcome intolerance and xenophobia. Our shared evolutionary origin implies deep genetic affinity among all people, but can this kinship and the prospect of reciprocity defeat the urges that divide us and magnify hostility? This new era has incentives to pollute and destroy without awareness; habits that nurture the surroundings and care for other species are antidotes to such inclinations. Intensification of our moral principles and the activation of meaningful values will guide the future path of compensation against the biases that degrade our surroundings or drown us in a sorrow of loss.

We now find repugnant that bigoted violence, slavery, and annihilation were ever deemed well-reasoned acts and expectations. Yet we must be repulsed, not because people are naturally disinclined to impose tragedy on others but rather because the meaning we discern in our fellow human beings, and in ourselves, has been transformed. What we come to value at the core of our lives is altered by reaction to the things that ultimately sicken us. In this experience lies hope.

Societies are jerry-built on the unforeseen outcomes of history. Future societies incarnating the Age of Humans are only now beginning to germinate. Each decade will be different from the one before it, as the human instincts to alter and to preserve continue their strange dance. We possess passions inclined toward sustaining the world as we know it, yet we live in a world that ever stretches us into the unfamiliar. Sustainability without adaptability is tomorrow's dead end.

Thus we return to the core question of human origin, now for a species dependent upon meaningful experiences of the world. Is it possible to transform, with informed purpose and care, our ways of living on this human-altered planet? Our responses will reveal the ever-evolving nature of what it will mean to be human.

❈ RETHINKING ECONOMIC GROWTH
PAULA CABALLERO AND CARTER J. BRANDON

Humans have gradually but relentlessly altered planetary life-support systems. Over time, as human progress has been increasingly predicated on continued economic growth (which has been measured almost exclusively since the mid-twentieth century by gross domestic product, or GDP), such growth has become an end in itself. Economic growth has helped to reduce poverty: when analysts reported that the share of the world's population living in extreme poverty had fallen from 37.1 percent in 1990 to a historic low of 9.6 percent in 2015, World Bank Group president Jim Yong Kim hailed the milestone as "the best news in the world today," adding that it marked real progress on the road to ending extreme poverty by 2030.

Yet economic growth and poverty reduction have come at the expense of the world's natural capital and ecosystems, as well as human health. The past decades have witnessed notable acceleration in resource degradation, climate change, pollution, and extinction. In fact, in 75 percent of all countries, the negative costs of environmental degradation are increasing right along with—and often far more rapidly than—economic growth. One terrible impact of such degradation is that almost nine million people die every year because of complications from air, land, and water pollution, according to the World Health Organization. Economic growth also has accelerated losses in the planet's biological diversity, with human activity contributing to both "the sixth extinction" of species and the third episode of global coral bleaching in the past 450 million years.

The reality of the Anthropocene is that a small percentage of the world's population has achieved a demonstrable and secure measure of prosperity and, in a very few cases, unfathomable wealth. However, the gains of the bottom 40 percent are insecure, and there will be challenges in meeting the expectations of the rising global middle class. In 2014, the World Economic Forum ranked widening income disparities as the second-greatest worldwide risk. The prevailing economic proposition has been that continued growth, through the steady distribution of benefits, would improve the lot of the poor and that concomitant environmental damages were an unfortunate, but not untenable, cost. Yet economists now recognize that the delivery of prosperity has been highly uneven. The relentless degradation of the world's natural assets, combined with the rising temperatures brought about by climate change, means that maintaining growth is no longer viable with business-as-usual approaches. The propositions of environmental sustainability and social equity require a major reevaluation.

The Anthropocene calls for radical changes in how we interact with nature, how we tackle poverty, and how we define prosperity. Given these challenges, the past few years have been momentous for the international development agenda. The definitions of new, *universal* frameworks that recognize the limits of planetary systems have been essential. Both the 2030 Agenda for Sustainable Development, approved by the United Nations General Assembly in September 2015, and the Paris Climate Change Agreement, approved in December 2015, acknowledge that all countries have a role to play in changing the prevailing development paradigm. These international frameworks recognize that sustainable—and inclusive—development must be more robustly defined, implemented, and monitored. While there is no silver bullet, certain tools and approaches will be decisive.

The Sustainable Development Goals (SDGs) adopted by the United Nations General Assembly acknowledge the necessity of breaking down the sectors (e.g., industry, transport, housing) that define how our economies, governments, and societies operate. Traditional economic measures of income and growth do not adequately capture or quantify what governments must manage to achieve long-term sustainability: since they reflect only monetary values, they do not reflect the need for clean air, clean water, and healthy ecosystems, especially when these are all threatened. The Nationally Determined Contributions (NDCs) that 163 countries pledged at the Paris Climate Change negotiations represent the beginning of thinking about economic growth in new ways while achieving the environmental objectives of reducing carbon emissions and enhancing climate resilience.

Today we can define sustainable development better than ever before, and new tools and methodologies are increasingly being applied to detect trends and calculate the economic and ecological losses that growth is inflicting on the planet. For example, we are learning to complement traditional national economic accounts with measures of produced, social, human, and, most important, natural capital. Integrating natural capital into national accounts helps countries measure the true value of that natural capital. It also helps them figure out the distributional impacts of ecosystem changes, such as who might enjoy the gains from investing in a healthier environment and who would ultimately bear the costs of ecosystem loss.

Investments in natural capital can generate high rates of return, whether measured by economic growth, total welfare, or targeted poverty reduction. In China, for example, environmentally targeted investments in the degraded Loess Plateau, a region the size of Belgium, have generated decisive benefits in land and ecosystem restoration, improved waterways, heightened food security, and increased employment. These investments also have doubled local incomes and lifted 2.5 million people out of poverty. In the United States, the benefit-to-cost ratio of investments to reduce air pollution under the Clean Air Act, estimated to be 25:1 in 2015, is predicted to be 31:1 by 2020. Other examples abound, from Brazil to Ethiopia to India.

Indeed, improved sustainability can be a forceful source of economic growth and poverty reduction, and not an impediment to either. A key challenge and opportunity of the Anthropocene will be advancing the SDGs on sustainable consumption and production. Promoting energy and other resource efficiency has been repeatedly shown to save production costs and improve long-term profits. Innovation in environmental technologies creates new markets, generates new jobs, and increases the productivity of labor and assets. Investments that reduce risk from climate-related extreme events also reduce costs by removing uncertainty from the marketplace. Finally, investments that reduce the impact of pollution improve the health of populations at large, enhance educational performance, and increase productivity.

Halting or reversing the negative trends of the Anthropocene will require us to change development trajectories, not just deliver immediate development outcomes. Low-carbon, resource-efficient, resilient, and inclusive growth demands continuing engagement in implementation and monitoring. We urgently need to take short-term actions to protect our long-term welfare, some of which involve investment (e.g., restoring degraded lands), while others require reforming policies (e.g., carbon pricing), improving governance (e.g., combatting illegal logging), or strengthening regulation (e.g., to reduce industrial pollution and overfishing). But we must also acknowledge within decision-making and monitoring processes that many investments in natural capital do not generate immediate returns, just as the benefits of education are not immediate but rather accrue to society over the following ten to fifty years. The short-term perspective that contributed to the creation of this new epoch needs to be overcome in favor of a long-term, holistic viewpoint.

Developing countries are growing faster than developed ones. In 2000, the developed countries constituted 60 percent of the world's economy (valued in terms of purchasing power parity), yet by 2030, that ratio will be reversed in favor of the developing countries. It is our fundamental responsibility to help ensure that this growth is not temporary but rather sustainable in economic and environmental terms.

At this point in the Anthropocene, hope can be found in the emerging transformation of the awareness, mind-sets, and behaviors that have defined the age's genesis. The SDGs and NDCs signal global agreement on the need to work together to reverse untenable trends. In many key areas, the private sector is leading the way, such as by ramping up renewable energy and reversing deforestation. Technology is revolutionizing human endeavors, including opportunities to support poverty-eradication efforts. Social media are redefining the contours of effective and informed participation, helping many advocate for greater accountability and transparency. The open question is whether our species can bring about the needed transformations at the scale and speed required to keep the planet within manageable parameters habitable to life as we know it.

The cumulative impacts of human activities—from agriculture, urbanization, and industrialization to the insatiable consumption of potable water, fossil fuels, forests, and other resources—have conjoined during the past ten thousand years to become drivers of planetary change, affecting the physical and biological state of Earth's terrestrial and marine environments as well as the chemical composition and physical dynamics of the atmosphere. These impacts are expected to intensify as the rate of change increases. The mere trickle of carbon dioxide that returned to the atmosphere when humans built their first wood fires has morphed into a rampaging flood of pollutants released through the massive combustion of the fossil fuels that power our everyday lives. Of course, fossil fuel use varies widely among the world's peoples, meaning that not everyone is equally culpable for this environmental scourge. The psychological dimension of this behavior adds another complication, as humans can embrace either their interdependence with the natural world or the illusion that they are not reliant upon Earth's finite resources. Increased atmospheric concentrations of carbon dioxide, methane, sulfur and nitrogen oxides, and aerosols have accelerated the green-house warming of the planet, which threatens to drastically alter the climate in the immediate and long-term future. Habitat degradation, also associated with global warming, is dramatically expressed in the world's forests, which have long been prized for conversion into croplands and pasture. Human migrations have produced more subtle—if no less profound—ecological impacts: people have been moving across oceanic and terrestrial environments for centuries, bringing along their favorite (and in some cases not so favorite) plants, animals, and microbes. Many of these transplanted organisms have thrived in their new, far-away habitats, becoming what some call invasive. This is especially true in urban settings, where native vegetation can be largely replaced by exotics. Yet while humanity's footprint is unmistakable in the world's burgeoning cities, it now extends to even the most remote, empty, and alien location imaginable—today hundreds of thousands of human-made objects circle Earth in near outer space. Expended rockets, dead satellites, and a multiplicity of other "space junk" litter the region enveloping our living planet. Human impact on every corner of Earth, and even beyond, is staggering. How we adapt and modify our behavior and activities will go a long way toward determining our destiny in the Anthropocene.

❊ THE FIRE THAT MADE THE FUTURE
STEPHEN J. PYNE

I sell here, Sir, what all the world desires to have—Power.

—MATTHEW BOULTON (1776)

In ancient times, fire was everywhere in daily life and, so it seemed, in nature as well. Heracleitus grandly asserted that all things were an exchange for fire, and fire for all things. Pliny the Elder, in tracing the means by which human artifice had transmuted raw nature into second nature, identified the critical catalyst as fire. Concluding his survey of the ways in which human intelligence could call upon art to help counterfeit nature, he marveled at the fact that fire was necessary for almost every operation. Fire was the keystone process for much of nature, and humanity the keystone species for how fire appeared in built and natural landscapes. Humanity's environmental power was a firepower.

So it made sense that creation stories for fire should serve double duty as creation stories for humanity. But fire did not come easily. It had to be taken through guile, force, or theft from ruling powers. It came with constraints: not every spark would kindle a flame, and not every flame could spread, unless its setting permitted. When Aeschylus wrote *Prometheus Bound*, about the Titan who brought fire to humanity, he had the hero boast that he had created all the arts of men, along with giving them hope. Fire made possible humanity's future. But Prometheus was chained to a rock in the Caucasus Mountains and punished for his deed. So too the fire he carried was restrained by the texture of the larger world.

• • •

That Promethean fire, encoded into cooking, crafts, hunting, foraging, farming, and herding, prevailed for millennia. Then fire on Earth underwent the greatest phase change since control over ignition was mastered by early hominins. It moved into machines, and the machines proved so ravenous that fire's tenders had to find additional fuels. They began to burn

40

lithic landscapes rather than living ones. The more fuel that was burned, the more they found. Revealingly, Thomas Newcomen's first steam engines were invented to drain coal mines. The emerging power began to feed on itself in the kind of autocatalytic process that fire had long epitomized.

The technological deconstruction of flame was under way. Gradually, working fires receded from human awareness as new technology found surrogates less prone to smoke and escape. Conceptually, too, fire ebbed from natural science as fire's core reaction became a subset of oxygen chemistry, fire's heat and light became derived phenomena of thermodynamics and electromagnetism, and animal heat yielded to the soft technology of metabolic physiology.

The architects of what they freely described as "fire engines" were celebrated as New Prometheans. In 1820, nearly twenty years after the invention of the first practical steamship and locomotive, Percy Bysshe Shelley published *Prometheus Unbound*, a celebration of the Titan as romantic, an unrepentant and defiant breaker of chains. The new fire too was no longer shackled by the ancient ecological fetters that had governed what burned, when it burned, and how it spread. The new fires could burn day and night, winter and summer, through drought and deluge, through ice age and interglacial. They were limited only by human ingenuity in finding new combustibles, and people directed much of their new firepower to just this quest.

A new fire was making a new future, as people took materials out of the deep past and projected them into what promised to be a very deep future. Until recent centuries, fire history could be broadly understood as a subset of natural history, particularly climate history. Increasingly, however, natural history, including climate history, must be understood as a subset of fire history.

• • •

Fire is a reaction, not a substance. It synthesizes its surroundings. So while scientists are fond of discussing processes in terms of drivers, fire more resembles a driverless car, no single set of hands on the wheel, barreling down the road and integrating everything around it. The same might hold for considerations of fire as a driver itself, but it's hard not to identify the new combustion as the power behind the Anthropocene. Humanity might be twisting the dials and pulling the levers of global change, but industrial combustion is turning the gears.

It is not just greenhouse gases that are the primary outcome, though their production promises to become another autocatalytic process of its own. It's the total effluent being dumped into ocean and biosphere. It's the fires that no longer burn living landscapes and are unhinging many biotas. And it's how people live on and organize the land. We move ourselves and parts of the landscape that we value by internal combustion: we use our unbounded firepower to reorganize whole continents. Indigenous species lost, invasive species gained, surviving biotas reassembled—all follow the lines of fire laid down by the new order of combustion.

This transformation in earthly combustion—what has been termed the "pyric transition," by analogy to the better-known "demographic transition"—typically begins with a population explosion of burning that ends with fire numbers below replacement value, unable to do the ecological work required. Historically, the fire problem humans faced was finding enough stuff to burn in the right way. Now the issue is coping with all the effluent released. Probably sometime in the 1970s, Earth began experiencing more emissions from the burning of fossil fuels than from burning landscapes. In a way familiar to fire history, that transition in burning has begun to multiply itself such that climate change is prompting greater landscape fires.

Cause, consequence, catalyst—fire can force change, fire can result from change, fire can assist change in ways seemingly removed from its immediate effects, and these changes can then propagate. In reshaping the order of fire, we have gone from remaking landscapes to remaking the planet. The fires of hearth and forge have both miniaturized into pistons and grown monstrous into dynamos. Previously tame landscape fires have gone feral. Once self-regulated wildland fires have gone rabid. In trying to confine fire too closely within our built world, we have let it run amok in the natural one, and now that unshackled combustion threatens to bind us with climatic and biotic fetters that could last for centuries.

• • •

Hesiod, Plato, Aeschylus—each told his own version of the Prometheus myth, adapted to his place and purpose. The moment has come to tell another version, better suited to our times and needs. To the good, the inherited story tells of our bond with burning, but it speaks to technological power abstracted, typically by violence, from its native setting and held in defiance of an existing order. Neither the unbound burning nor the defiant hubris can be tolerated much longer.

We need a new myth—a creation story for the Anthropocene. The fire of the future will not be a source of raw power but a companion on our journey, part of our stewardship of the planet, a power we bind and bend to a greater biotic good. We need a Prometheus who is not a fire thief but a fire tender, a Prometheus who is not a disruptor but a guardian, a Prometheus willing to subject himself to shackles because unchained he is a danger to all, not least himself. We need fires not just to make a future but to sustain one.

✻ A NEW DREAM OF THE EARTH
WADE DAVIS

The term *Anthropocene* suggests that we have entered an era in which human beings, empowered with a transformative authority equal to the evolutionary and geologic forces of nature, will determine the destiny of the planet. As an organizing principle, the concept sends a powerful message to a species fully capable of adapting to any degree of ecological degradation. But as a metaphor, the notion is flawed, implying as it does that the environmental crisis is a consequence of the actions of humanity as a whole. Nothing could be further from the truth.

To be sure, our species has never had a purely benign presence on the planet. Our Paleolithic ancestors hunted wildlife to extinction. Plant and animal domestication during the Neolithic only accelerated the erosion of the wild as the agricultural frontier spread. But the truly devastating human impact on the environment, the sequence of events that led to this moment, occurred in but fifteen human generations reaching back to the beginning of the Industrial Revolution. The entire history of the European colonial encounter reminds us that most human cultures played little, if any, role in the development and implementation of the ideas and technologies that ultimately resulted in an all-out assault on the natural world.

The crisis of the Anthropocene was in fact provoked by a relatively small subset of humanity, a specific and unique cultural tradition that ultimately reduced the world to a mechanism, the planet to a commodity, with nature seen as but an obstacle to overcome. It is not humanity that has brought on the crisis, but rather a set of beliefs and attitudes that are most assuredly not held by the vast majority of people with whom we of the mechanized realm share the planet.

All cultures are ethnocentric, fiercely loyal to their own interpretations of reality. The names of many indigenous societies translate as "the people," the implication being that every other human is a nonperson, a savage from beyond the realm of the civilized. We too are culturally myopic and often forget that we represent not the absolute wave of history but merely a worldview, and that modernity is but an expression of our cultural values. It is not some objective force removed from the constraints of culture. And it is certainly not the true and only pulse of history. It is merely a constellation of beliefs and economic paradigms that represent one way of doing things, of going about the complex process of organizing human activities.

It's useful to recall where these ideas originated. During the Renaissance and well into the Enlightenment, in our quest for personal freedom, we in the European tradition liberated the human mind from the tyranny of absolute faith, even as we freed the individual from the collective, which was the sociological equivalent of splitting the atom. In doing so, we also abandoned many of our intuitions around myth, magic, mysticism, and, perhaps most important, metaphor. The universe, René Descartes declared in the seventeenth century, is composed of only mind and mechanism. With a single phrase, all sentient creatures aside from human beings were devitalized, as was Earth itself. Science, as Saul Bellow suggested, made a housecleaning of belief. Phenomena that could not be positively observed and measured could not exist. The triumph of secular materialism became the conceit of modernity. The notions that land could have anima, that the flight of a hawk might have meaning, that beliefs of the spirit could have true resonance were dismissed as ridiculous.

For several centuries the rational mind has been ascendant, even though science, its finest expression, can still in all its brilliance answer only the question "How?" but never come close to addressing the ultimate question: "Why?" The inherent limitations of the scientific model have long provoked a certain existential dilemma, familiar to many of us taught since childhood that the universe can be understood only as the random action of minute atomic particles spinning and interacting in space. But more significant, the reduction of the world to a mechanism, with nature but an obstacle to overcome, a resource to be exploited, has in good measure determined the manner in which our cultural tradition has blindly interacted with the living planet.

I was raised, on the coast of British Columbia, to believe that the rain forests exist to be cut. This was the essence of the ideology of scientific forestry that I studied in school and practiced in the woods as a logger. My cultural perspective was profoundly different from that of the First Nations, including those living on Vancouver Island at the time of European contact and their descendants who are still there. If I was sent into the forest to cut it down, a Kwakwaka'wakw youth of similar age was traditionally dispatched during his Hamatsa initiation into that same forest to confront Huxwhukw and the Crooked Beak of Heaven—cannibal spirits who live at the north end of the world—with the goal of returning triumphant to the potlatch as a fully socialized human being infused with spiritual discipline and fortitude. The point is not to ask or suggest which perspective is right or wrong. Is the forest mere cellulose and board feet? Was it truly the domain of spirits? Is a mountain a sacred place? Does a river really follow the ancestral path of an anaconda? Who is to say? Ultimately, these are not the important questions.

What matters is the potency of a belief, the manner in which a conviction plays out in the day-to-day lives of a people, for in a very real sense this determines the ecological footprint of a culture, the impact that any society has on its environment. A child raised to believe that a mountain is the abode of a protective spirit will be a profoundly different human being than a youth brought up to believe that a mountain is an inert mass of rock ready to be mined. A Kwakwaka'wakw boy raised to revere the coastal forests as a realm of the divine will be a different person than a Canadian child taught that such forests are destined to be logged. The full measure of a culture embraces both the actions of its people and the quality of their aspirations, which are revealed by the nature of the metaphors that propel them onward.

Herein, perhaps, lies the essence of the relationship between many indigenous peoples and the natural world. Life amid the malarial swamps of New Guinea, the chilly winds of Tibet, or the white heat of the Sahara leaves little room for sentiment. Nostalgia is not a trait commonly associated with the Inuit. Before the arrival of industrial logging, nomadic hunters and gatherers in Borneo had little overt sense of stewardship of mountain forests that they lacked the technical capacity to destroy. What these cultures have done, however, is to forge through time and ritual a relationship to Earth that is based not only on deep attachment to the land but also on a far more subtle intuition—the idea that the land itself is breathed into being by human consciousness. Mountains, rivers, and forests are not perceived as inanimate, as mere props on a stage upon which the human drama unfolds. For these societies, the land is alive, a dynamic force to be embraced and transformed by the human imagination.

As we move forward, it behooves us to listen to the voices of the many hundreds of cultures struggling to be part of the global dialogue that will define the future of life on Earth. There are currently fifteen hundred languages gathered around the campfire of the Internet, and the number is increasing by the week. These indigenous voices matter because they can still remind us that there are indeed alternatives, other ways of orienting human beings in social, spiritual, and ecological space. The point is not to naïvely suggest that we attempt to mimic the ways of nonindustrial societies, or to ask any culture to forfeit its right to benefit from the genius of technology. It is, rather, to draw inspiration and comfort from the facts that the path we have taken is not the only one available and that our destiny is therefore not indelibly written in a set of choices that demonstrably and scientifically have proved not to be fully wise. By their very existence, the diverse cultures of the world bear witness to the folly of those who say that we cannot change, as we all know we must change the fundamental way that we inhabit the planet.

❋ LOCATING OURSELVES IN RELATION TO THE NATURAL WORLD

LINDSAY L. CLARKSON

A certain way of understanding human life and activity has gone awry, to the serious detriment of the world around us.

—POPE FRANCIS, *LAUDATO SI': ENCYCLICAL LETTER OF THE HOLY FATHER FRANCIS ON CARE FOR OUR COMMON HOME* (2015)

In his *Laudato si'*, Pope Francis reviews the current state of human impact on Earth, linking our treatment of the environment to the uneasy state of our internal worlds. He highlights the spiritual alienation and the pervasive sense of emotional impoverishment in our acquisitive society. He calls on us to pay attention to and take responsibility for our interaction with the natural world: "to become painfully aware, to dare to turn what is happening to the world into our own personal suffering and thus discover what each of us can do about it."

The pope's proposition is familiar territory to the discipline of psychoanalysis, which is uniquely concerned with how our internal worlds affect our recognition of and interaction with external reality. A central premise of psychoanalysis is that unconscious fantasies about the nature of the relationships we carry within us serve to organize our engagements with other people and with the natural world. Although we would prefer to think that we interact logically and rationally, in reality we are pressured and influenced by emotionally charged fantasies of "the way things are."

All awareness and aliveness is accompanied not only by well-being and satisfaction but also by pain, envy of others, and frustrations caused by limitations inherent in the perception of dependence, separateness, and mortality. We all must encounter and try to cope with such facts of life. There are two fundamentally different ways to react to the problems intrinsic to human awareness: toleration and evasion. We can bear with the frustrations and losses that are a part of being alive, or we can make every effort to destroy our knowledge of the situation that leads to pain.

Psychoanalysts locate the primitive origins of the divergence in means of psychological coping with disappointment in the complexity of mother-and-infant interactions. From the infant's point of view, one would think that being held, fed, and nurtured would be ideal and sustaining. But to be dependent on another person means acknowledging that one is not omnipotent and not in charge of limitless resources. Someone else has a bounty not possessed by oneself. Envy and a feeling of humiliating smallness can ensue, interfering with a good feeding relationship between the baby and the mother. The fact that the person on whom one depends has finite attention and resources, including a restricted life-span, may arouse such a terror that the awareness itself may feel unbearable.

Human dependence on the natural environment evokes similar anxieties. We can be primarily in tune with the natural world, receiving from nature, as the pope's encyclical advises, "what nature itself allowed, as if from its own hand." In this frame of mind, there is an acceptance of the human place in the web of life, which encompasses knowledge of biological complexity, the passage of time, and mortality. There is a comfort and a secure home in knowledge of ourselves as part of a larger natural system that we cannot control but nevertheless depend upon. However, if dependency is equated with humiliation, we reject such a place and seek dominion over the environment. We take from the natural world ruthlessly and discard as rubbish what is not immediately useful or monetarily valuable, with an omnipotent fantasy of limitless resources and no bounds to growth. Because of our projected aggression, the environmental situation can become menacing in our minds, leading to a cycle of dangerous amplification.

We are perpetually in internal conflict: should we embrace and safeguard our interdependence with the wider natural world or turn toward illusion, disavowing our dependence on Earth's finite resources, acting as though anything we wish were possible? In today's world, societal forces encourage us to be distracted by acquisition rather than to find solace in living in reality, tolerating the perennial complexity of balancing restraint and satisfaction. A secure, primarily benevolent relationship to our internal objects allows tolerance of worry and frustration, and may shift the balance toward bearing a sense of responsibility for our common environment and toward taking action on its behalf.

Two poems by W. S. Merwin illustrate these different intrapsychic positions with respect to nature. Merwin is a poet of exquisite sensitivity to the natural environment and its vulnerability to human impact and interference. Evoking the beauty and continuing process of nature, he is able to situate human existence, with its complexity of aggressiveness, love, loss, and mourning, within the cycles of the natural world.

> *To the Insects*
>
> Elders
>
> we have been here so short a time
> and we pretend that we have invented memory
>
> we have forgotten what it is like to be you
> who do not remember us
>
> we remember imagining that what survived us
> would be like us
>
> and would remember the world as it appears to us
> but it will be your eyes that fill with light
>
> we kill you again and again
> we turn into you
>
> eating the forests
> eating the earth and water
>
> and dying of them
> departing from ourselves
>
> leaving you the morning
> in its antiquity

Here Merwin is speaking to the insects, to us fellow human sojourners, and to himself in a thought experiment that probes the mind-set of someone who finds our links to animal life unbearable. The narrator—a first-person plural human voice that addresses the insects as "elders"—describes a state of mind in which humans have set themselves apart from nature and, by doing so, depart from themselves. He alludes to the losses inherent in alienating ourselves from the community of life in order to create beliefs to protect ourselves from the awareness of the natural limits of our existence. Knowledge of reality is evaded through distortion of memory, dislocation from the natural world, and the fantasy that ravenous consumption and destruction can continue without consequences. The narrator intimates that he is aware of his evasion of knowledge but cannot face up to truth in a sustained way: "departing from ourselves // leaving you the morning / in its antiquity."

Merwin describes a bleak end to human existence. No comfort prevails in the mechanical, persecutory take-over by insects. The view is paranoid: insects in this poem are not objects of love or objects of appreciation for the intricacies of their lives and the important and complex roles they play in interrelationships between growth and decay. The light in their eyes is devoid of warmth. We imagine the narrator: lonely, disconnected, hunkered down in fear or aggressively denying that limits to his desires exist. Such a state of mind would not lead to environmental concern.

Twenty years later, Merwin draws us into a different position:

The Laughing Thrush

O nameless joy of the morning

tumbling upward note by note out of the night
and the hush of the dark valley
and out of whatever has not been there

song unquestioning and unbounded
yes this is the place and the one time
in the whole of before and after
with all of memory waking into it

and the lost visages that hover
around the edge of sleep
constant and clear
and the words that lately have fallen silent

to surface among the phrases of some future
if there is a future

here is where they all sing the first daylight
whether or not there is anyone listening

Immediately the reader is struck by the warmer, fuller emotional tone of
this poem. The poet is a listener, appreciative at a musical level. He is describing a moment in time, the song "tumbling upward" from the darkness. The
darkness is not unfriendly; it is the source of richness. The bird's song reflects
a way of being that is unlike that of a person, who is always questioning and
bounded by individual self-consciousness. The narrator expresses tolerance
for the unknown, the hush, the dark valley. Instead of antagonism or irrita-
bility, we are situated in a deep sense of time ongoing: "in the whole of before
and after / with all of memory waking into it." The listener is awakened to all
of the reverberations in his or her own memories, moved by conscious and
unconscious stirrings. In a reverie, the narrator is aware of "the lost visages."
He is receptive, letting them come. The visages are not at all vague. The
strength of the images in the narrator's mind's eye implies a sturdy relationship
to a good internal object. In this state of mind, primitive anxieties are diminished. With an appreciation of the fullness of life, a tolerance for mortality,
and a relaxation into the continuity of nonhuman life, one can imagine such
a person roused to action to preserve the natural world.

These two poems by Merwin clearly illustrate the opposing states of mind
that coexist within each of us in an ongoing equilibrium. How we interact
with the natural world depends on the inner emotional climate we inhabit
most of the time. The preponderance of human effects on Earth at the
present time reflect the ruthless, dominating, and omnipotent aspects of our
nature. But with encouragement and a sense of belonging, we can face up
to the anxieties stirred by our ancient dependence on Earth's resources and
by our part in the damage already inflicted on the environment in which we
live. From this position, we can discover what each of us might contribute to
a more benign and reparative engagement with the natural world.

❧ TEMPERATE FORESTS
A TALE OF THE ANTHROPOCENE
SEAN M. McMAHON

A fascinating feature of folk stories is the emergence of similar narrative themes in diverse cultures, independent of language or location. These commonalities flow from the universal ability of stories to impart lessons, offering morality tales, cautions about dangers, and guides for life's transitions. The Anthropocene has defined a new era of Earth's story, as humans have ceased to be simply characters in and instead have become formidable writers of this planet's tale. Forests, for example, feature in many folktales and yet can be seen as their own tale. Spanning millennia and cultures—from medieval Europe through imperial China to the development of a global timber trade—humans have consistently told a similar story through their relationship to forests, from timber removal to forest management and from fear through conquest to concern. The tale of temperate forests exemplifies the way in which humans have confronted and been forced to understand and reconsider the biological world that they have so aggressively disrupted and yet into which they have always been integrated.

For thousands of years, humans have drastically altered the terrestrial world between the tropics and the polar circles. Despite a limited growing season, sufficient rainfall and summer temperatures across these regions permit the growth of highly productive forests that contain some of the largest organisms on Earth. The remaining colossal redwood and sequoia groves in California, the stems of *Eucalypus regnans* in Australia that rise to more than three hundred feet, and photographs from the 1920s of giant American chestnut trees in the preblight years are clear examples of the incredible potential for exuberant life in these systems. Yet the same rich soils that support these forests are also part of the arable land that allowed the development of human civilization.

Because good land for trees is often good land for crops and domesticated animals, forests around the world compete with crops and animal husbandry, two fundamental requirements of subsistence living everywhere. A third requirement is shelter, and the trees felled to make way for food can be used as materials for habitation or for fire. It takes little imagination to envision how forested regions of the temperate world were repeatedly and persistently invaded over the past several millennia by human settlements, small scars within intact forested landscapes, appearing and expanding along rivers and trails. From the perspective of these small colonies, however, forests were more than just an obstacle to subsistence. They also invoked a sense of menace. They were dark and forbidding and served as a refuge for very real threats. These woods, which were everywhere between towns, were considered wilderness, and not in the contemporary sense of "pure" or "natural" but in King Henry IV's meaning when he says, in Shakespeare's *Henry IV, Part 2*, "O my poor kingdom . . . ! / O thou wilt be a wilderness again, / Peopled with wolves, thy old inhabitants!" Wilderness is the antithesis of settled civilized life, and to some, forests exemplified this opposite. The benefits, then, of clearing forests are land for pasture or crops, wood for structures and fires, and the security and almost moral virtue of "taming" a threatening wild.

As economies developed, however, new patterns emerged in the story of humans and forests. Whether in China in the fourth century, Western Europe after the Renaissance, or New England in the seventeenth century, forests shifted from an obstacle to settlement to a resource for civilization. Naval dominance by Britain required ships, and ships required wood—up to six thousand mature oak trees for a single ship. With England largely deforested, the American colonies became a critical source of timber. The iron industry's need for wood-derived charcoal, which began as early as the fifteenth century, stressed the remnant forests of Europe and eventually affected forests throughout the world. The loss of forest cover led the expanding industries of the nineteenth century to switch from charcoal to fossil fuels such as coal. In Asia, population increases in the premodern era matched accelerated declines in remnant forests. As in Europe, the percentage of forested land remaining at the start of the Industrial Revolution was due to variations in the landscape or the culture: forests remained in locations that were difficult to access, holy, or preserved for elites (such as for hunting) or had slopes too steep for farming, grazing, or easy timber harvest. When the good timber was gone, the remaining trees were used as pulp for paper or charcoal for stoves. Regardless of the different times when these geographically distant societies changed their use of the forests they controlled, the reduction of those forests to their lowest extents converged around the middle of the nineteenth century. Gradually, these societies began to understand that the resource they had taken for granted was being exhausted. Attention to the regrowth and management of forests was needed.

The result of this acknowledgment of the limits of forests was somewhat heartening. In many parts of the temperate world, a substantial release of land from agriculture led to a resurrection of forests, as has been documented in England, France, China, Japan, and eastern North America. Land records and, later, aerial photographs show entire regions where forest stands emerged as cohorts. In these stands it is easy to see how species composition and structure reveal imprints of dynamic land-use changes and population growth. An example is found in New England, 98 percent of which was forested before Europeans arrived. By 1840, after the harvesting of timber for ship masts and charcoal, the expansion of colonization, and the rise of American industries, only 25 percent of these forests remained. Today forests cover approximately 75 percent of New England, but only a fraction of these new forests are due to natural regeneration. Similarly, by 1800 some form of silviculture had been established in both Asia and Europe, and by 1900 timber management was a global project of resource production, just as agriculture had been in previous centuries.

Management of timber can take many forms. The practice of clear-cutting stands to replace them with a single fast-growing species has maximized the use of timber material in managed forests but also has made them vulnerable to pest outbreaks and climate shifts. A new emphasis on longer-term sustainability is growing in importance across countries with temperate forests. The Menominee nation in Wisconsin has long practiced timber production in this way, managing their forests for hundreds of years with a "seven generation" approach. They plant, cultivate, and remove trees across a variety of landscapes and through different climate patterns, considering both profitability and sustainability. Their practices include single-tree harvesting; cultivating a combination of species that best corresponds to given soils, topography, and past forest composition; harvesting damaged trees; and leaving healthy trees in longer rotations than most commercial timber companies do. Such sustainable practices are now being recognized as critical to global temperate forest management and afforestation initiatives worldwide, especially as climate change and invasive pests and pathogens seriously threaten species-poor and hastily harvested forests.

Scientific research on forest growth and health has been instrumental in the recovery of temperate forests over the past century, as accurate descriptions of the ecological dynamics of forests have informed effective efforts to preserve and regrow them. Temperate forests were the first to have large monitoring plots installed; in Czechia, for example, careful measurements of tree location, size, and species date back to the late nineteenth century. Such monitoring programs can follow the dynamics of long-lived and complex systems. The Smithsonian Institution's Forest Global Earth Observatory (ForestGEO) is a network of research plots where every stem is identified, mapped, and measured every five years. ForestGEO has extensive coverage across China, North America, and Europe, with large plots demonstrating the influence of land-use history, species diversity, and climate on the productivity and health of temperate forests. These research forests are now highly instrumented—everything from photosynthesis in the upper canopy to the chemical links between roots and soil microbes may be measured. These efforts are directed at learning how forests evolved to their current state and then using that knowledge to better predict how they may change in the coming decades.

A folktale may teach a lesson about life or society. The lesson of the tale of temperate forests is clear: to live sustainably, humans must consider how best to preserve, regrow, manage, and protect forests. Many societies have learned this lesson, but not all have changed their practices to reflect the knowledge gained. Tropical forests are now being treated in a way similar to that which led to the vast disappearance of temperate forests; further, the deforestation of the tropics is largely a response to the demands of the temperate zone, undertaken for the economic benefit of its societies, which need only look into their own histories to understand what the fate of the tropics may be. In a fast-changing world that will put ever more demands on land and resources, a global plan is needed for the implementation of management policy and practices that stem from a scientific understanding of how forests function. Initiatives to manage, monitor, and sustain forests have emerged as important features of local and global conservation efforts, but there remains a distressing disconnect between the clear arc of the tale of temperate forests and the path that society is on.

❊ URBAN NATURE / HUMAN NATURE
PETER DEL TREDICI

From their humble beginnings as isolated settlements at the intersections of important transportation routes, today's cities have evolved into pillars of the Anthropocene. They house half of the world's human population and produce 70 percent of its fossil fuel emissions while occupying only 3 percent of Earth's land area. In cities, human values—driven by socioeconomic factors—trump ecological values, such that people encourage the presence of organisms that make the environment a more attractive, livable, or profitable place to be, and vilify as weeds and pests those species that flourish in opposition to these goals.

In most modern cities, the native vegetation that originally occupied the site is long gone. In its place, one typically finds a cosmopolitan array of species—some planted intentionally, some growing spontaneously—that are adapted to the ecological conditions generated by the city itself. In urban areas in northeastern North America, it has been estimated that 25 to 40 percent of the spontaneous vegetation is nonnative, a figure that rises to 70 percent when one looks at only the densely populated core regions. Just as they are for people, cities are melting pots for plants, and questions about where they came from become irrelevant after a few generations.

From a functional perspective, most vegetated urban land can be classified into one of three broad categories: remnant native landscapes, which are left over from the earliest days of urban settlement and are composed mainly of native plants; managed horticultural landscapes, which are composed of horticultural plants cultivated for specific purposes (e.g., ball fields, parks, gardens, street trees); and abandoned or neglected landscapes, which no one takes care of and which are dominated by plants that flourish without human intervention. Depending on the socioeconomic conditions of a given city, this last land-use category can make up from 5 to 40 percent of the total area. While most people have a negative view of spontaneous urban plants, they are actually performing many of the same ecological functions that native species perform in nonurban areas. In short, they help make cities more livable by absorbing excess nutrients that accumulate in wetlands; reducing heat buildup in heavily paved areas; controlling erosion along rivers and streams; mitigating soil, water, and air pollution; providing food and habitat for wildlife; and converting the carbon dioxide produced by the burning of fossil fuels into biomass.

Cities display a suite of environmental characteristics not typically found in natural habitats. The most significant is the ongoing physical disruption and land fragmentation associated with the construction and maintenance of infrastructure. Ongoing construction destabilizes native plant communities by altering soil and drainage conditions, thereby creating opportunities for the establishment of disturbance-adapted, early successional plants. Going hand in hand with disturbance is the covering of most urban land with pavement and structures that shed water. The density of these impervious surfaces is greatest in the center of the city and decreases as one moves out to the edges, while the amount of open ground typically follows the reverse pattern. Compounding the problem of imperviousness is the issue of soil compaction produced by pedestrian and vehicular traffic. This reduced porosity inhibits the flow of air and water into the soil and can be particularly serious for native trees and shrubs whose shallow roots require a constant supply of oxygen for proper growth. When the pavement in a town exceeds 25 to 30 percent, it can be considered urbanized from the biological perspective, independent of the density of its human population.

In most cities, the quality of the soil, like that of the vegetation, is a mixed bag. One can certainly find existing pockets of native soil that support remnant native ecosystems, but most cities, especially those along coasts, have large areas filled in with construction rubble to create more land. Roughly 17 percent of Boston as it currently exists is built on fill soil, as are significant parts of New York City near the ocean. Such filled land, by definition, can never support a native ecosystem—which is not to say that it cannot support a functional cosmopolitan ecology.

Another distinguishing characteristic of urban environments, and a function of their abundance of impervious surfaces, is their high temperatures relative to the surrounding nonurbanized land, a phenomenon referred to as the urban heat island effect. Because buildings and pavement absorb and retain heat during the day—to say nothing of cars, air conditioners, heating units, and electrical equipment, which also generate heat—the annual mean temperatures of large urban areas can be between 2 and 5°F (1–3°C) higher than those of the surrounding nonurban areas. On cloudless summer nights, the temperature difference between the center of a large city and the nearby countryside can be as much as 18°F (10°C). This means that cities are already providing people with a preview of what climate change will look like on a much broader scale in the not-too-distant future.

Based on extensive research in Europe, scientists have determined that the typical urban plant is well adapted to soils that are relatively fertile, dry, unshaded, and alkaline. Through a twist of evolutionary fate, many of these species have evolved life-history traits in their native habitats that have "preadapted" them to flourish in cities. Marble or brick buildings, for example, are analogous to naturally occurring limestone cliffs. Similarly, the increased use of deicing salts along walkways and highways has resulted in the development of high-pH microhabitats that are often colonized by either grassland species adapted to limestone soils or salt-loving plants from coastal habitats. Finally, the hotter, drier conditions one finds in cities favor species that come from exposed, sunny habitats in nature. Preadaptation is a useful idea for understanding the emergent ecology of cities because it helps to explain why some plants and not others grow on piles of construction rubble, chain-link fence lines, highway median strips, pavement cracks, and compacted turf.

Any discussion of urban ecology would be incomplete without a consideration of the cultural significance of the plants that grow in cities. Indeed, the changing composition of spontaneous urban vegetation over time reflects the constantly shifting value judgments, socioeconomic cycles, and technological advances that shape the evolution of cities. The shift from horses to automobiles in the early twentieth century, for example, effectively transformed land once used for hayfields and pastures into roads and parking lots, thereby reducing available habitat for many grassland plants and animals.

While most biologists view invasive plants as a serious biological problem, the fact remains that their initial introduction and distribution were usually the result of deliberate decisions that reflected the economic, ornamental, or conservation values of the day. Between the 1930s and the 1960s, various federal, state, and local agencies encouraged—and often subsidized—the cultivation of plants such as kudzu, multiflora rose, and autumn olive for erosion control and wildlife habitat purposes. It should come as no surprise that they became major problems forty years later, after millions of them had been planted. Indeed, the spread of nonnative species across the landscape is as much a cultural as a biological phenomenon, a fact often overlooked by advocates of strict ecological restoration.

The interacting forces of urbanization, globalization, and climate change have led to the formation of novel associations of plants that have become the de facto native vegetation of the city. These plants not only reflect the city's socioeconomic history but also project its future trajectory. Given the environmental uncertainty that is a hallmark of the Anthropocene, now is the time for people to acknowledge the role that spontaneous urban vegetation can play in helping to clean up the ecological mess that we have made of the planet.

❈ ATMOSPHERICS AND THE ANTHROPOCENE
KELLY CHANCE

On a lovely September night in 1885, Samuel Pierpont Langley measured the Moon's light throughout a lunar eclipse. Those infrared measurements and Langley's research on the temperature of the Moon and on solar heat provided the physical chemist Svante Arrhenius with data that led him to propose the theory of greenhouse warming of Earth. Published in 1896, Arrhenius's theory includes an early mathematical formulation of the atmospheric greenhouse law that is still useful today.

• • •

Greenhouse warming. As Arrhenius predicted, global warming is real, and it is here. The prime physical cause of warming is that a blanket (in this case, a blanket of infrared-absorbing gases) retains warmth. Denying global warming means invoking feedback effects that are as large as the primary warming. Carbon dioxide (CO_2), known in Langley's and Arrhenius's time as carbonic acid, is the main contributor to this blanket. It has reached an atmospheric concentration of 400 parts per million, up from about 280 in preindustrial times. Methane (CH_4) is next in importance: its atmospheric abundance has almost tripled since preindustrial times. Modern methods of extracting natural gas, which is mostly methane, cause still unquantified amounts of leakage, and gas and oil wells that are no longer in production may continue to be significant sources of methane. Synthetic chlorofluorocarbons (Freons) remaining since their phase-out, after it was determined that they cause stratospheric ozone depletion, and their hydrochlorofluorocarbon and hydrofluorocarbon replacements are smaller but significant greenhouse sources, as is nitrous oxide (N_2O), which is about 40 percent anthropogenic in origin and is growing in concentration because of fertilizer use.

Global warming produces much more than a gradual rise in temperature. Thanks to the thermal expansion of seawater and glacial melting, sea levels are rising. Circulation changes induced by warming also cause anomalous weather, including droughts, flooding, and increased storm frequency.

Air pollution. The day following Langley's eclipse measurements, London experienced its earliest snowfall on record, heralding the beginning of an extremely long, snowy, and smoggy winter. The city was infamous for its pea-soup fogs, which occurred when emissions from the burning of low-grade coal, notably inexpensive and plentiful sea coal, combined with ambient fog, usually in humid conditions with temperatures below 5°C. These noxious fogs had already been a problem for centuries when John Evelyn produced his report *Fumifugium: Or, the Inconveniencie of the Aer and Smoak of London Dissipated, Together with Some Remedies Humbly Proposed* for King Charles II in 1661. They still infested London in 1905, when the term *smog* was coined for them, and they culminated in the four-day Great Smog of December 1952, when twelve thousand people are estimated to have died as a direct result of this pollution.

London fog is an exemplar of "reducing-type" pollution: it is both highly sulfurous (sulfur dioxide, SO_2, is produced in abundance by burning dirty coal) and high in particulates such as coal ash and sulfuric acid. Los Angeles smog is another exemplar: it is oxidizing pollution, or photochemical smog, produced by the action of sunlight on nitrogen oxides emitted from combustion, and it is rich in ozone (O_3) and liquid-phase aerosols. Global urban pollution varies in type between these extremes and with pronounced seasonality, and it may include significant particulate sources such as desert dust or black carbon from forest fires and other biomass burning. But tropospheric pollution is not confined to urban areas. Given the ubiquity of emissions and the transport of ozone and aerosol precursors, unhealthy concentrations can occur over extended rural areas. Nitrogen oxides from combustion (in power plants, for example) and volatile organic compounds are the major rural tropospheric ozone sources. Forests stressed by heat—and perhaps by ozone itself, in a pernicious feedback loop—can produce prodigious amounts of ozone and aerosols.

Tropospheric ozone, nitrogen oxides, sulfur oxides, aerosols, carbon monoxide, and lead are the U.S. Environmental Protection Agency's criteria pollutants. Among them, ozone and aerosols cause the most atmospheric concern globally at present because they have the most serious health consequences. Aerosols with particulate matter whose diameter is 2.5 microns or less ($PM_{2.5}$) are of particular concern for cardiovascular disease and respiratory health. Ozone is especially harmful to people with asthma or heart disease and may cause or exacerbate respiratory illness. Forest and crop damage by ozone is well documented and serious: it retards photosynthesis, sometimes leading to the "early autumn" syndrome, and makes some plants more susceptible to disease.

The ozone layer. Both the *Iliad* and the *Odyssey* mention the smell of lightning-produced ozone. By the mid-nineteenth century, unstable, highly explosive ozone gas had been isolated. The stratospheric ozone layer was discovered in 1913 and was soon being measured in detail on an ongoing basis. It is described by the 1930 Chapman chemical mechanism involving oxygen species and sunlight, with the later addition of water-vapor-initiated ozone-reduction chemistry. The ozone layer shields life in the lower atmosphere, where pollution and global warming predominate, from chemical bond–breaking ultraviolet radiation, which is particularly harmful to animals and plants. Not until the early 1970s did a handful of brilliant scientific studies establish that the ozone layer is subject to damage from anthropogenic activities, especially production of the precursors of free-radical nitrogen and halogen chemical sources (NO_x and ClO_x). Also beginning in the 1970s, ground-based measurements by the British Antarctic Survey showed rapid, deep declines in stratospheric ozone during austral spring. This discovery was confirmed by National Aeronautics and Space Administration measurements by the Nimbus 7 satellite, which mapped the extent of the depletion. It was found to correspond to the Antarctic polar vortex. Satellites have continuously mapped global ozone and the Antarctic ozone hole ever since. Intensive investigations, including expeditions to Antarctica, have shown that stratospheric halogen chemistry—especially affected by chlorofluorocarbons, particularly Freons, whose use has now been largely stopped—created the ozone hole.

• • •

Together the three phenomena described above form the canon of atmospheric consequences of anthropogenic activities: greenhouse warming affects heat and circulation; atmospheric pollution causes chemical and physical changes to the air; and depletion of the stratospheric ozone layer allows increased surface ultraviolet radiation. They are often conflated, but they are distinct, albeit deeply intertwined. The major feedback mechanism is that increased greenhouse warming increases pollution. Droughts, for instance, increase forest fires, which increase pollutants. Heat increases power usage (to run air conditioners, for example), which also increases pollution, although increasing nighttime and winter temperatures can reduce power use for domestic heating. Heat also increases stress on vegetation, which increases volatile organic compound production (of isoprene and related compounds), which increases ground-level ozone production. However, aerosol pollution generally acts to cool the climate. Greenhouse warming may also exacerbate stratospheric ozone depletion by odd hydrogen (HO_x) chemistry as increased convection injects water vapor into the stratosphere; decreased stratospheric ozone, however, increases tropospheric ultraviolet radiation, which acts photochemically to reduce tropospheric methane, a greenhouse gas. Pollution may increase greenhouse warming by ozone in the lower atmosphere.

Drawing conclusions is more difficult than framing questions, as the future depends heavily on human capacity and political action, but questions must be asked. Will denial of anthropogenic consequences fade? Will plans to lessen emissions work in the face of a growing population? Will feedbacks of greenhouse carbon dioxide and methane from land reservoirs accelerate? Most important, will alternate energy sources prove economical and effective? A positive answer to this last question may be our greatest hope for mitigating both global greenhouse warming and atmospheric pollution.

❧ BEYOND THE BIOSPHERE
EXPANDING THE LIMITS OF THE HUMAN WORLD
LISA RUTH RAND

The idea of the Anthropocene as a potential geologic epoch indicates the global scale of human influence on the geophysical world. Although the official start date of the Anthropocene continues to be a topic of controversy among scientists, some measurable changes to the planet simply cannot be chalked up to natural variations in the Holocene. Perhaps the most compelling—and incontrovertible—hallmark of this proposed new epoch circles hundreds, thousands, and even millions of kilometers overhead and out of sight. Beginning in the mid-twentieth century, a small club of spacefaring nations used new technologies to launch satellites and other anthropogenic objects into the nearest reaches of outer space, extending the spatial dimensions of human-driven change outward. As the anthropologist Alice Gorman has argued, the Anthropocene cannot truly be understood without considering the place of outer space in geophysical change.

Recent scholarship in environmental history has connected the pronounced acceleration of global environmental change during the Cold War to the shifting geopolitics and environmental movements of the era. During the first decades of the Cold War space race, the United States and the Soviet Union shaped the true wilderness of outer space into a landscape: Beginning with the launch of Sputnik in the autumn of 1957, the reach of *anthropos* expanded beyond the tenuous boundaries of the biosphere. Although the nearest reaches of outer space are popularly portrayed as an empty void—the opposite of the green-and-blue nature portrayed by environmentalists and artists alike—in truth they support an abiotic ecosystem defined by energy exchanges, radioactivity, natural rocky objects and energetic plasmas, and gravity variations. Regular solar cycles drive the "weather" of near-Earth space, which is home to most artificial satellites and uncontrolled orbiting waste, colloquially known as space junk. However, the climate in space and climates on Earth are not separately impacted by human activity, and terrestrial climate change has affected the near-Earth ecosystem: Increased carbon dioxide emissions during the Great Acceleration have caused the planet's thermosphere to contract, reducing the altitude at which atmospheric drag brings space junk out of orbit and diminishing the resilience of this ecosystem. Indeed, these exchanges suggest an interactive continuum of natural and anthropogenic exchange between Earth and space environments in what the anthropologist Valerie Olson calls "the extended ecological heliosphere."

By 1961, some 380 trackable anthropogenic objects circled the planet. By the end of 1963, that number had mushroomed to 685—not including debris too small to be detected by the space surveillance technology of the time. At most recent count, the Space Surveillance Network estimates that nearly seventeen thousand objects large enough to be tracked orbit the planet. Of these, 77 percent have been confirmed as space junk—objects with no designated use or purpose. Another approximately five hundred thousand pieces of debris between one and ten centimeters (0.4 and 4 inches) in diameter and more than one hundred million anthropogenic particles smaller than one centimeter make up the system of artificial waste in near-Earth space. These minute, largely invisible bits of anthropogenic material orbit alongside the empty rocket bodies, dead satellites, and other large objects speeding overhead. Awareness of this accumulating debris drove the first debates, in the late 1950s, over what counts as pollution in outer space.

As soon as the first satellites had reached orbit, both the United States and the Soviet Union sought to use this new environment for both peaceful and nefarious purposes. High-altitude and exoatmospheric nuclear weapons tests began in 1958 and ended in 1963 with the Partial Test Ban Treaty. New forms of satellites—from giant, shiny inflatable balloons to a ring of hundreds of millions of tiny copper fibers—tested the use of space for communications while spurring controversy over whether they could interfere with astronomy, crowd the electromagnetic spectrum, or present a collision hazard to other spacecraft. Current geoengineering proposals intended to correct atmospheric change by "seeding" orbital space with reflective material recall some of these early controversies and could spur similar debate over the environmental protection of near-Earth space. As outer space became a site of technological utility, scientists around the world came to embrace Earth orbit as a valuable site of investigation and rapidly abandoned the idea of a physically isolated Earth.

As the Space Age wore on, the larger objects became a threat—not just to other objects in orbit but to people, property, and environments on the ground. The natural geophysical ecosystem of the Earth-Sun environment accelerated into overdrive during the peaks of the first eleven-year solar cycle to take place after the launch of Sputnik. As the atmosphere expanded, uncontrolled objects succumbed to friction and fell back to Earth through the destructive upper atmosphere. With no control over where surviving fragments might land, orbital space became a site from which pollutants could cross geographic boundaries and extraterritorial regions. In cases such as that of the nuclear-powered satellite Cosmos 954, which fell over the Northwest Territories of Canada in 1978, reentry raised the very real specter of radioactive contamination of ground, sea, and sky—with little more than the caprice of the natural geophysical forces of Sun and space to blame. From nuclear weapons tests to the reentry of radioactive debris, the markers of the nuclear Great Acceleration extended from underground to outer space.

Many current space-policy analysts claim that orbital debris is a recent problem—a result of the ignorance or negligence of the spacefaring nations of the 1960s. However, this argument ignores the early transnational discourse on environmental risk that began with the first attempts to measure, understand, and use the near-Earth space environment. As the first pieces of space junk circled the planet, newly minted space scientists reached the consensus that the physical influence of our planet extends dozens of kilometers into a physically interactive solar system. Just as the Great Acceleration has coincided with the rise of mainstream environmentalist movements around the world, concern over environmental risk was extended into the nearest reaches of outer space from the first moments of the Space Age. The satellite and probe technologies that humankind launches into space both expand our understanding of the universe and make up an information infrastructure that supports modern globalism as we currently know it. Should the orbital environment change to the point that it can no longer sustain the satellite infrastructure, many large technological systems would fail, with repercussions for industries from communications and finance to agriculture and disaster relief. Bruno Latour has suggested even further consequences should a space debris crisis occur—perhaps permanently closing humankind off from the rest of the cosmos, without an extraterrestrial future should we render the biosphere uninhabitable.

As difficult as it may be to visualize human-driven global change, thinking about the Anthropocene also requires shaking up a more basic, seemingly innate set of assumptions about the world around us. It requires us to reconsider our ideas about nature and the natural and redraw the limits of the human environment to include even those places most alien, most empty, most seemingly antithetical to verdant nature as we know it. The Earth of the Anthropocene is a much larger world than can be bounded by biology, geology, or atmosphere. Through the production, use, and discarding of spacecraft and space junk, humanity has broadened the boundaries of anthropogenic geophysical change into the universe, rendering our species truly, if virtually, cosmopolitan.

III
RESPONDING

TO CHANGE

In the Anthropocene, societies will face an array of unfamiliar and often stressful environmental conditions, many of which their cultural, political, and economic systems were never designed to address. Moreover, the impacts of global environmental change will not be equally distributed across the planet. Future responses to this change may be novel or may incorporate lessons drawn from prior adaptations to climate fluctuations and other environmental upheavals. Archaeologists have documented humankind's long history of deliberately (and often unintentionally) modifying our surroundings in response to change. The world's indigenous peoples, many of whom occupy lands hard hit by the early impacts of global warming, have much to say about adaptive practices. Groups and individuals who have been displaced to foreign lands are no less concerned about their immediate environments than those who remain on their native soil. In the United States, the insidious and often subtle influences of racism have hampered the effectiveness of society's pursuit of environmental protection by defining the cause as a white concern, thereby limiting the development of broader and stronger coalitions. The mounting stresses that threaten the resiliency of the world's forested landscapes highlight the necessity of cooperative action infused with scientific understanding, as exemplified by the movement to adopt a socioecological approach to human intervention in forest succession. Agriculture constitutes the single greatest use of land by humans, who have been "gardening the planet" for millennia, and thus developing sustainable agricultural and horticultural responses to climate change is a paramount challenge as communities strive to provide sufficient food, fuel, and shelter for rapidly expanding populations. Similar interventions will be necessary in marine habitats, as anthropogenic environmental change imperils them, too. Planetary transformations will also have significant consequences for human health, both physical and psychological, especially in urbanized areas, where negative impacts are often magnified. As human populations increase to numbers never reached before, not only will we face the threat of new and possibly more virulent infectious diseases, but unprecedented efforts will be required to build healthy lifestyles in an ever more crowded world.

ARCHAEOLOGY AND THE
FUTURE OF OUR PLANET

TORBEN C. RICK

Where and how do humans fit into the natural world? These questions have long puzzled philosophers, historians, artists, biologists, and others who have sought to understand our place on the planet and beyond it. The massive challenges that face Earth's biodiversity and ecosystems today give these questions added relevance, especially as people grapple with an uncertain future in the Age of Humans.

The question of where humans fit into the natural world guides the research of many archaeologists. Archaeology provides an important perspective on basically everything that makes us human, from where our species came from to the origins of hierarchy, written language, economics, animal and plant domestication, and agriculture. It also provides insight into how people have interacted with their environments over centuries and millennia, evidence of which is often found in the plant and animal remains that are preserved in the archaeological record and the information we glean from them through identification and quantification, including genetic, isotopic, and other technical analyses.

Archaeologists have long focused on understanding how past fluctuations in climate and other environmental changes have influenced human societies. We have also increasingly explored the ways that humans in the past altered and modified their environments—both intentionally and unintentionally. The archaeological record abounds with examples of the successes and failures of peoples over millennia of occupation and evolution. The Classic Maya in Mesoamerica are a prime example: some two thousand years ago, they were an impressive and sophisticated society with monumental architecture, social hierarchy, and complicated political and ritual systems. Although debate rages about the precise causes of their decline, a combination of persistent drought, overexploitation of resources, and deforestation likely contributed to the dramatic reorganization or collapse of Maya polities.

Several publications, both scientific and popular, have explored the ways in which the Maya, the Easter Islanders, the Southwest Anasazi, and other groups altered their environments, often using these examples as cautionary tales for our own potential fate. Some of the popular accounts are oversimplistic, painting a picture of humans as always overusing resources or as perpetually shortsighted. While we cannot deny this scenario, the archaeological record also contains many remarkable success stories for human societies—such as the persistence of the Maya for centuries. Archaeology can help us to understand the major issues of the Anthropocene through such cautionary tales of success and failure and can provide direct links to contemporary conservation and management. Bones, shells, and other animal and plant remains, which offer clues to how past peoples managed, altered, or depleted resources, also allow us to understand long-term environmental change and better prepare for changes forecast for the next few decades to centuries.

Many archaeologists study hunter-gatherers, who are not generally known for the massive landscape and ecological changes of early states and more complex groups such as the Egyptians, Maya, and Aztecs. Still, much remains to be learned from their archaeological record, as is illustrated by knowledge generated over the past ten years that has helped to contextualize modern-day environmental problems. Shell middens—an archaeological site type containing the remains of all the trash that people left behind, usually fish, mammal, and bird bones, shells, plant remains, and artifacts—and other archaeological sites serve as vast repositories that shed light on past human-environmental interactions across space and over long periods of time. Analysis of the animal and plant remains documents which types of organisms lived in the past, whether humans had an impact on their abundance, and how human activities and climate change affected prehistoric patterns of biotic distribution. Like snapshots of the past, these data can be compared to modern-day patterns to show how we arrived at the present moment and how ecosystems have changed through time.

Such comparisons can help us to establish baselines or targets for conservation and restoration. For instance, we know that some species are declining today, many are near extinction, and we are living through our planet's sixth mass extinction. To understand and transcend this situation, we need to know how we got here. What do species' histories tell us about the composition of natural communities? Which environmental and cultural targets are we trying to achieve? Why do we choose to conserve the organisms and ecosystems that we do? All such questions come down to desired future conditions—that is, what do we want the ecosystems of the future to look like? Will they be places of great abundance and high biodiversity? Will they represent what environments were like before humans arrived (which would mean going back well over one hundred thousand years ago in some cases)? Determining desired future conditions is a key challenge and goal of the Age of Humans. It is also one that should involve a wide cross-section of society, from the public to policymakers to scientists. Archaeologists, as a group uniquely positioned to document long-term human-environmental interactions, potential sustainability, and the interplay of human activities and climate change, will have an important role to play.

Interdisciplinary teams of researchers and scholars including archaeologists, biologists, ecologists, geneticists, historians, paleobiologists, and many more will help us to answer these questions. One example of such a team investigated the ecosystems, organisms, and people of the Channel Islands, five of which compose a national park off the coast of California. This project used archaeological and fossil samples, modern biological specimens, and genetics to understand species of conservation concern, such as the island fox (*Urocyon littoralis*). Its work has unraveled the history of the fox and other species, both with and without the presence of people and under a range of climatic conditions, and the data now serve as benchmarks for the management plans of the National Park Service and the Nature Conservancy. This investigation is a potent example of success at the local level;

however, climate change and the other challenges of the Age of Humans are global problems.

In reality, we need both global and local approaches. Success on the local level can give hope, serve as an example of what to do elsewhere, and motivate others to positive future action. Local-level research, however, must be done in concert with global, policy-driven discussions such as those among nation-states working to combat climate change. Collectively, all these efforts indicate that we can transcend the challenges of the Anthropocene. Although archaeology shows our past pitfalls, it also gives us reasons to be optimistic by revealing how truly ingenious we humans can be. The key now is to make a global call to action that begins at the local level, is driven by interdisciplinary research, and uses our planet's past to help guide its future.

❦ LIVING ON A CHANGING PLANET
WHY INDIGENOUS VOICES MATTER

IGOR KRUPNIK

By the 1980s, mounting scientific evidence pointed to global climate change as one of humanity's most daunting problems. The subsequent international debate often drew attention to particularly vulnerable areas of the planet—small islands, high mountains, tropical forests, low-lying coastal areas, and polar regions. Besides their heightened sensitivity to the early impacts of climate change, many of these areas have been homes to distinct groups of indigenous peoples.

In spite of this disproportionate attention to their homelands—which were often undiplomatically called the canaries in the coal mine of global climate change—indigenous groups have had to fight hard to gain a place in the high-level international discussions of climate change. Yet their perspectives have proved more welcome in certain areas than in others. For example, because Earth's northern polar region has experienced one of the globe's most pronounced shifts in its climate regime over the past several decades, the voices of indigenous Arctic residents were heard loud and clear by 2000, with their observations and concerns about the changing climate and weather widely reported. As the case of the Arctic demonstrates, a strong consensus among indigenous peoples, scientists, politicians, and governments on the threats brought about by the changing environment and the need for collective action is possible.

Similar, albeit slower moves to recognize the value of indigenous knowledge of climate change have taken place in small-island habitats, mountainous areas, tropical forests, arid lands, and other regions populated by indigenous peoples. In the past fifteen years, as the dams of political neglect

---

This essay draws on ideas first presented by the author at the United Nations University–United Nations Educational, Scientific and Cultural Organization conference "Indigenous Peoples, Marginalized Populations and Climate Change," held in Mexico City in July 2011.

and sidelining have been breached, scholarly papers, international conferences, research initiatives, special journal issues, and books on indigenous peoples and global change have proliferated.

The lasting value of indigenous people's voices in the climate change debate is based on four messages that come from their specific knowledge, observations, and practical solutions in the changing environment. These messages may be critical to strengthening the new intellectual framework that is emerging to address global change.

THE MESSAGE OF LOCAL SCALE

From a climate science perspective, present-day climate change is a global process. Yet people commonly experience its impacts in local contexts. The critical role of indigenous peoples in anchoring or scaling down global change scenarios is often overlooked. As a large portion of the world's population celebrates the new global village, indigenous people continue to live locally—in the context of the habitat they have longed called home. It is the piece of the planet they know best and observe daily. They keep a record of its minute changes, which they pass from generation to generation. Every study on indigenous peoples and global climate change confirms the richness and the value of indigenous observations and interpretations, which international organizations and government agencies are increasingly willing to accept.

By 2020, climate scientists and environmental agencies will be forced to translate their global and regional scenarios into high-confidence local models and plans. When this happens, the knowledge and high-resolution vision of indigenous peoples may offer an authoritative template. We may argue that scientists and policymakers will look to indigenous peoples not as much for new data as for a general philosophy to interpret the data they already possess. Such a philosophy will focus on intimate features of individual habitats, local adaptations, grassroots initiatives, attention to emotional and spiritual well-being, and growing self-reliance—all trademark features of indigenous peoples' knowledge systems.

THE MESSAGE OF SELF-RELIANCE

Indigenous peoples historically viewed themselves as being responsible for the health of their habitats in a practical, social, or spiritual sense—usually all three—and, in most cases, they continue to do so. Environmental scientists increasingly call such an approach ecosystem stewardship, and many argue that it should be promoted to the level of planetary stewardship.

Emerging from many recent studies of indigenous adaptations is a recognition that indigenous peoples are able to keep their "houses" (meaning their habitats) in order. They rightly claim that they possess a thorough knowledge of their environment and have maintained sustainable links with their habitats via technological and spiritual means over generations—rotating crops and plots, following nomadic routes, maintaining local networks of exchange, and even singing and drumming for the animals. Far from passive victims of change, they are actively observing, experimenting, and evaluating alternative livelihood strategies, just as they have always done.

Their individual stories coalesce into a powerful message of self-reliance, contrary to the common assumption that indigenous peoples are mere casualties of global change because of their small numbers or simple technologies. This message promises to gain strength as the world's attention shifts from planetary modeling toward addressing specific threats in specific locales through the development of local-scale initiatives and grassroots solutions.

THE MESSAGE OF INDIGENOUS PEOPLES' RIGHTS

Indigenous peoples have been instrumental in adding a human rights agenda to the global climate change debate. Documents including those generated at the 1992 United Nations Earth Summit indirectly addressed this topic through discussions of economic inequality, colonialism, and conflicts between the world's industrial and developing nations. However, the specific aspect of minority and human rights has been conspicuously absent from many high-level discussions of global change. There are reasons to look at climate change and indigenous peoples through the lens of social justice—or rather, injustice. The question is straightforward: how can we respond to climate change in ways that are just?

Climate change has disproportionally affected many of the world's indigenous groups, whose situation has been made worse by the historical legacy of unjust treatment within their nation-states. Their lands are now mere fractions of their former tribal habitats, as these groups were shunted to regions with extreme environments—areas with limited water resources or at high elevations or in remote locations that others found undesirable. Indigenous peoples commonly have lower income levels and fewer resources for adapting to change than their nonindigenous neighbors in the same society, while they are subjected to webs of regulations by the management agencies charged with administering their lands and affairs.

Indigenous peoples' demand for environmental justice also articulates the plights of other disadvantaged and marginalized groups, such as small sharecroppers and fishers, women and children, and urban and rural poor, who often have no representation in international bodies. As indigenous peoples argue for recognition of their traditional knowledge, respect for their use of land, and acceptance of their right to membership in climate change agreements, their voices at the table help to expand our common ethical and societal sensitivity. Their engagement also puts a human face on the international climate change negotiation process, which is often domi-nated by government-to-government politics and ideology and conflicting national claims.

THE MESSAGE OF ACTION

As the world's representative bodies struggle to establish meaningful strate-gies to combat global climate change, indigenous peoples' message of self-reliance increasingly stands as a call for action. Almost every study of indigenous communities' adaptations to environmental change under-scores the proactive nature of their responses to the threats associated with a warming planet.

Some of their actions are surprisingly akin to strategies being applied by other local players, such as small municipalities, town mayors, nongovernmental organizations (NGOs), and citizen groups, that are increasingly taking matters into their own hands by promoting alternative energy sources, reliance on green technologies, horizontal networking, resource sharing, and reduced pressure on vulnerable ecosystems. It is almost certain that the growing connections among indigenous peoples, scientists, NGOs, and other grassroots programs will provide a viable alternative to the top-down practices of national governments and international agencies. Architects of such alternative approaches may draw on many strengths inherent to indigenous people, such as intimate knowledge of local environments, cultural resilience, kinship and regional networking, flexible economies, and close bonding with nature.

One way or another, the politics of global climate change will shift. A new era will begin that ushers in a growing role for local voices everywhere—in integrative assessments, grassroots self-reliance, recognition of human rights in climate change adaptations, and collective actions to combat environmental threats. This transition promises to frame the international climate change debate for years to come, and the growing role of indigenous people in this process is hard to overestimate.

✳ BLACK AND GREEN
THE FORGOTTEN COMMITMENT
TO SUSTAINABILITY
LONNIE G. BUNCH III

In the spring of 1971, four African American freshmen from Howard University decided to attend the second Earth Day celebration on the National Mall in Washington, DC. Walking south from the campus, they soon joined a small army of college kids from around the nation, all bent on reminding America that protecting the planet should matter to all who inhabit it. The day unfolded like so many other gatherings of the period, with speeches, chants directed at Richard Nixon, and music by groups such as the Beach Boys. The Howard students sat behind a group of other students from an array of colleges, who suddenly noticed their presence. One woman asked them, with a mixture of questioning and confrontation, "What are you black guys doing here? This isn't a civil rights demonstration."

The notion that issues of environmentalism are outside the scope, interest, and historical concern of black America is both a misreading of American history and a failure to grasp how often lax laws and environmental change have shaped, damaged, and affected communities of color. The fallacy that grappling with the environment is solely a white concern has limited the development of alliances with African Americans, which would strengthen the coalitions needed to battle for the protection of the environment.

African Americans, who have often borne the brunt of environmental injustice, have a long and complicated history of protecting their communities while combatting environmental pollution. In almost any American city, the most environmentally at-risk areas are frequently in neighborhoods of poor or working-class communities of color. Whether from mercury-soaked lots in South Central Los Angeles, tainted landfills near Newark, New Jersey, or the drinking water in Flint, Michigan, recently discovered to be unsafe, African Americans have felt the greatest impact of the United States' willingness to compromise environmental protection. How many historically black communities, such as those in Saint Paul, Chicago, and Providence, Rhode Island, have been divided and negatively impacted by the construction of interstate highways, which have fractured communities and adversely affected air quality? These impacts are not accidents but part of considered decisions that sacrificed the health and cohesion of communities of color for the "greater good" of regional progress and economic growth. The disastrous 1927 flood in Mississippi and the devastating 2005 inundation of New Orleans following Hurricane Katrina demonstrated that black communities and black lives are not as valued or protected as those of white America when environmental calamities strike.

Yet African Americans have not just been victims of environmental abuse and neglect. They also have a long history of involvement in sustainability movements. Until the 1920s, the majority of African Americans lived in the South and worked the land, some as owners of small farms, many as sharecroppers or tenant farmers. All were dependent upon the land for their survival. The novelist John Williams described this connection by simply stating, "Black people are land people." Limited resources and reliance on agriculture for their survival conditioned many African Americans to find ways to protect the land and to develop lifestyles based on sustainability.

These farm families were early advocates of recycling, the careful use of limited resources, taking a long-term view toward land use, and learning about, understanding, and adjusting to changing climate conditions—all of which were essential to the existence of the millions of African Americans who tilled the land. The work of historically black colleges often supported and developed these notions of sustainability. In the late nineteenth and early twentieth centuries, colleges such as the Tuskegee Institute in Alabama and the Hampton Institute in Virginia created rural cooperatives and agricultural extension services that brought the most recent agrarian knowledge and domestic efficiencies to black farm families. While these farmers could not define *sustainability* or *environmentalism*, their daily practices and values not only provided the means for survival but established traditions that benefited their land and the country.

Recognizing and acknowledging how African Americans have made America better—whether through a forgotten commitment to environmentalism, the struggle to find racial justice, or contributions that shaped American culture—is one of the goals of the newest museum of the Smithsonian Institution, the National Museum of African American History and Culture, created by congressional legislation in 2003. Millions of people who visit this inclusive institution will remember or learn about the rich, complex, and often unknown history of black America. While the museum helps America confront its tortured racial past—visitors can ponder the pain of slavery and segregation—it is also a place to find the joy and the strength that are so much a part of the African American experience. But if a museum only helps people to remember, then it is fulfilling only part of its mandate. This museum uses African American history as a lens to better understand what it means to be an American. It does not create a historical narrative for black people alone. The museum suggests that the resiliency, optimism, and spirituality at the heart of the African American community have shaped much of America's identity and that many of the moments when Americans embraced an expansion of liberty or a redefinition of citizenship were shaped by or emanated from within the African American experience. Ultimately, the vision for the museum is to provide a safe space where engaging exhibitions, scholarship, and educational and public programs enable Americans to grapple with the issue of race that has so divided us and with the goal of making America better.

It is ironic that when the National Museum of African American History and Culture decided on the characteristics that would shape the architectural design of its new edifice, one particular aspect evoked great surprise. While the building had to speak of a permanent African American presence and reflect spirituality and uplift, it was also essential that it be sustainable and green and obtain the standard of Leadership in Energy and Environmental Design Gold. Some of the initial comments from the public and Smithsonian colleagues were reminiscent of the young woman who could not understand why African Americans cared about the environment.

Why was sustainability so important? Would resources not be better spent on other aspects of the building's development? Did the Smithsonian have the expertise to oversee such a design? Regardless of the challenges, it was crucial that the building remind us of what is often overlooked. So much of African American history is unacknowledged or hidden in plain sight, and, since the history of black America is also tied to notions of sustainability, it was essential that the building be green. This was a way to recognize that the struggle to protect the planet, to find fair and equitable environmental treatment, is a civil rights issue, an issue that has had a major impact on black lives both historically and contemporarily. One hopes that this new, sustainable museum will encourage and stimulate other museums to follow suit, and that it will remind all who visit that race matters when it comes to environmental issues.

❧ FOREST SUCCESSION AND HUMAN AGENCY IN AN UNCERTAIN FUTURE
ROBIN L. CHAZDON

Forests, like all ecosystems, are constantly changing. As naturalists, foresters, scientists, and users of forest products, we impose our human constructs on spatially and temporally dynamic tree-based systems that defy static definitions. The forces that shape the composition, structure, and ecosystem properties of forests—geologic history, climate, natural disturbances, harvesting, fires, fragmentation, and agricultural land uses—act synergistically to influence the past, present, and future of the socioecological systems we call forests.

Uncertainty has been and always will be a major element in the regeneration of forests, and succession is nature's way of responding to both uncertainty and opportunity. The first individuals and species that arrive in a newly opened habitat play a decisive role in the unscripted play of forest succession. These pioneers can accelerate or inhibit future species colonization, which is also strongly affected by local conditions, fauna, and features of the surrounding landscape. These interacting factors have always been a part of the assembly processes that characterize regenerating forests, often leading to highly variable successional trajectories even within the same region. We are just beginning to understand the complex interactions among landscapes, fauna, and forest regeneration, which will inform how these factors can be manipulated to restore forests and landscapes. Our nascent understanding will need to expand quickly to meet the certain demands of an increasingly uncertain future.

Early ecologists viewed succession as a local, progressive, and orderly process of species replacement culminating in a stable climax community in equilibrium with the local climate. Today's ecologists hold a nonequilibrium view of succession, emphasizing the role of recurrent disturbances in shaping population and community structure. Currently, forest succession is considered a hierarchical process, incorporating effects of the surrounding landscape on the structure of local populations and communities. But ecological processes are still seen as the primary driver of successional dynamics, limiting humans to the role of external agents of disturbances that initiate succession (by, for example, harvesting plants and animals or clearing and burning vegetation) or alter successional trajectories. Species "foreign" to the original ecosystem are regarded as disturbance agents rather than as components of regenerating ecosystems. This view also holds that ecological knowledge is necessary and sufficient to understand, manage, and restore tropical forests.

Many forest scientists now challenge this perspective on environmental dynamics. We live in a complex world that does not conform to a strictly ecological paradigm of forest succession; instead, human agency must be integrated into the successional paradigm and not simply relegated as an external factor. Forest succession is a socioecological process and has been so wherever and whenever human populations have coexisted with forests.

Consider the dichotomies that have marked our understanding of forests and people: unmanaged versus managed, natural versus unnatural, intact versus fragmented, pristine versus disturbed, novel versus historical, conservation versus restoration, mitigation versus adaptation. These contrasts are increasingly blurring. Moreover, their forced distinctions impede the establishment of resilient forest landscapes, which form the foundation of Earth's life-support system. Now, with three-quarters of the terrestrial biosphere altered by human activity, concepts of native ecosystems and native landscapes are increasingly challenged and tested. Terms such as *anthromes* (anthropogenic biomes) and *neolandscapes* (anthropogenic landscapes) have enriched our vocabulary. In many parts of the world, the forests of the future are not going to resemble the forests of the past. Let's move on.

Expanding human populations are placing increased demands on the remaining forests, which are shrinking in extent, have lost key animal mutualists, and are impacted by multiple biotic and abiotic stressors. This unsustainable pattern can be seen across the planet. How can we expect the ecosystem functions of Brazil's once vast and unbroken Atlantic Forest to be performed by the small, isolated, and struggling forest patches that occupy less than 12 percent of its former expanse? Something has got to give.

Add to this mix projected scenarios of novel climates and increasingly extreme climatic events, as climate change is expected to bring about new combinations of temperature and rainfall that will test the adaptive capacity of species and directly affect human populations and geographic patterns of land use. The dynamics of intact forests, forest fragments, and newly established reforests—plantations, restored forests, agroforests, and everything in between—will be directly and indirectly impacted by interacting climate and anthropogenic stressors.

We need to help new forests establish, grow, and prosper, whether through active planting and careful management or through spontaneous or assisted natural regeneration. There are too many interdependent variables to allow predictions of how all these factors will interact and affect future forest dynamics and landscape change. But some general trends can be predicted. First, successional forests will be increasingly prevalent across the world as both natural and anthropogenic disturbances increase in frequency and intensity. In the aftermath of human migration and changing land-use practices, forest ecosystems will spontaneously regenerate whenever and wherever they are afforded the chance. We can encourage and create opportunities for this process.

Second, successional forests will be increasingly humanized and homogenized within mosaic landscapes and close to urban and agricultural areas. Increasingly, their composition will likely comprise both generalist species and those species adapted to disturbed habitats that are broadly distributed and have wide ecological tolerances. Nonnative species will increasingly predominate within and along the edges of these second-growth forests, as is already the case in the eastern United States, in subtropical regions of Queensland, Australia, and on islands such as Hawaii, Puerto Rico, and

Réunion. More diverse secondary forests composed largely of native species will be found in the buffer zones of protected areas in remote regions.

Third, successional forests will be increasingly important in both climate change adaptation and mitigation and will play critical roles in water regulation, soil stabilization, and nutrient conservation. They will also provide havens for biodiversity and create living corridors reestablishing landscape connectivity for wildlife, including links across different elevational zones, which will allow species to migrate to more favorable climatic conditions.

A socioecological paradigm of forest succession views human interventions as being inherent elements of disturbance regimes. Forest dynamics emerge from feedbacks from social and biophysical drivers that affect the functioning of ecological subsystems and determine the potential for sustainable management. Long-term monitoring of social and biophysical characteristics across different tropical landscapes is required to reveal the interplay of local, regional, and global drivers of successional forest change. Research conducted within an integrated socioecological framework is urgently needed to provide insight into critical issues such as the effects of global change on tropical ecosystems (including changes in climate, biodiversity, land use, and land cover), future scenarios for biodiversity conservation, sustainable provision of ecosystem goods and services, and options for sustainable forest-based livelihoods. Further, socioecological research can elucidate the drivers of tropical forest degradation and thus guide short- and long-term interventions to halt future degradation and restore forests on lands where regenerative capacity has been lost.

Restoring the world's forests will also restore humanity. Ensuring a promising future for forests requires that we go beyond recognizing and valuing the many goods and services that forests offer us. It is time to develop a real mutual partnership between people and forests. We must provide for the needs of forests, become stewards of forests, and help guide them through hard times. We need to recognize both their resilience and their vulnerability, and we need to begin now. The life-support systems of our planet are at stake.

❋ OCEAN 2.0

J. EMMETT DUFFY

When I was born, in 1960, the ocean was still a mysterious and often frightening world, full of watery wildernesses few had seen because only a handful of scientists and adventurers had yet breathed underwater. The deepest spot in the ocean was plumbed that very year, but most of the vast depths remained unknown, and even on the surface sailors traveled thousands of miles without seeing a trace of humanity. As the space age dawned, many still scoffed at the idea that the ocean could be depleted of fish.

Those days are gone. The high seas are still a wilderness of sorts, as we realize when supertankers are hijacked and disappear seemingly into thin air. But the revolutions in technology that the space age ushered in have filled in nearly all the blank regions on the map. Many of us now have experienced the magic of the underwater world, some by scuba diving, the rest by pressing a remote-control button from the comfort of their couch. Robots prowl the deep-sea floor in search of rare metals. And fish can no longer hide from the military-inspired technology that routinely brings fresh seafood to our tables from the far corners of the globe. Dizzying advances in many fields of science have opened the ocean's treasure chest and revealed its riches in more detail than was imaginable in the mid-twentieth century.

Most of us—more than seven in ten people on Earth—live within one hundred miles (160 kilometers) of a coast. But even in the continental interiors of Minneapolis and Mongolia, the unseen sea is central to human life and livelihoods. The ocean is one of Earth's two lungs, its microscopic algae producing half the oxygen we breathe. In the process, these plants act as a biological pump, absorbing much of our combusted fossil carbon and sequestering it in the deep ocean. Similarly, we have the ocean to thank for taking the heat for us, absorbing 93 percent of the warming produced by our industrial metabolism and reducing climate change to a fraction of what it would otherwise be. Sea life is a key source of humanity's protein, especially in the developing world—the average person on Earth eats twice as much fish as poultry and three times as much fish as beef. And the ocean connects us: sea shipping carries more than 90 percent of the trade generated by our appetites. If the ocean were a country, it would have the sixth-largest economy in the world.

But the ultimate sea change is now upon us. The signs are alarmingly familiar—declining fisheries, whirlpools of plastic debris, oxygen-starved dead zones, and more. So what are we going to do about it? Happily, there is light at the end of the tunnel, and if we can reach that source, it can guide us toward a soft landing in the brave new ocean. Finding and amplifying the bright spots will require deeper understanding both of the vast, interconnected ocean ecosystems that sustain us and of our own unique species, which now dominates their dynamics. In the past decade or so, the global community has recognized these challenges and risen to them with innovations in natural and social sciences, making progress on both fronts. Technological advances in particular have transformed environmental science and conservation in sometimes spectacular ways. One major impetus has been the satellite and geospatial technology that we now take for granted when navigating our cars to the soothing voice of Siri. Another is the flourishing of social media that allow scientists and members of the general public throughout the world to collaborate, advancing both innovation and democratic decision making. A striking example that unites these themes is the nonprofit SkyTruth's online sharing of satellite imagery showing what's happening in near-real time across the world's oceans. By making this data publicly available, SkyTruth's Global Fishing Watch program fosters crowdsourced identification of pirate fishing by illegal and disguised vessels—among the most pernicious threats to ocean life— and apprehension of the culprits. Crowdsourced satellite tracking has also shut down seismic surveys by a ship exploring for oil on the sensitive Mesoamerican Barrier Reef in violation of Belize law.

Satellites give us an unprecedented view of the ocean's surface, but they can't penetrate it. Many challenges still require boots on the ground—or fins in the water. Recognizing that there are too few professional marine biologists to cover the great expanses of the ocean, the Reef Life Survey program has engaged a large and enthusiastic community of recreational scuba divers, training them rigorously to census reef animals and habitats and logging surveys at more than four thousand sites worldwide to produce an unparalleled database on marine biodiversity. This coalition of scientist and citizen-scientist divers has gathered hard evidence documenting how marine ecosystems shift in response to climate change and showing that large reserves and enforced protection from fishing provide a major boost to ocean biodiversity.

The bright spots in ocean science and conservation are the result of purposeful actions, many of them necessitated by evidence of declining ecosystem services. But we can't fix problems if we don't see them in the first place. In our age of accelerating change, global real-time information is more important than ever to inform critical decisions. A special challenge is tracking the wildly diverse species and interactions that are the heart of functioning ecosystems. Building such a capacity requires innovations in both technology and the ways we interact with one another and the world around us. The Smithsonian's Marine Global Earth Observatory (Marine-GEO) program embodies that spirit, marshaling technological and social innovation together. It is building a networked community of scientists and volunteers around the world who combine next-generation DNA sequencing, drone-based habitat mapping, and other tools to create an open-access resource for understanding changing marine life and ecosystems.

Perhaps the brightest ray of hope for nature in the Anthropocene is that the current rapid pace of technological change is being paralleled by the rapid evolution of human attitudes in many areas—gender and race equality and animal welfare, among others. There are hints of a great acceleration in human awareness and engagement in safeguarding our planet. Environmental consciousness concerning the ocean took some time to awaken but has spread like wildfire in the age of social media. I am comforted by young people everywhere passionately wanting to change the world for the better and by the growing international agreement on climate action. Nearly one hundred countries have banned shark fishing or the sale of their fins, and many airlines now refuse to transport shark products. In 2015, more of Planet Earth was officially protected than in any other year in history, and most of that area—more than two million square miles (five million square kilometers)—was underwater.

Committing to a change in course that avoids massive disruption of human communities and economies is the central policy challenge of our era. Making that course correction will require harnessing and amplifying the passion and ingenuity already evident in current sustainability work. Although the Anthropocene ocean will be very different than the one we have known, there is still time to ensure that it will be healthy and productive. A fundamental transformation similar to that of the land surface has not yet happened in the sea, and the ocean's big animals are mostly still with us. We need above all to transition to a carbon-neutral economy. And we must recognize that most of the ocean is no longer wilderness—it requires wise spatial planning, just like the land. That planning will depend on an intimate knowledge of how the ocean's diverse and beautiful life-forms interact to create healthy ecosystems. We have a chance. But we don't have time to waste.

❊ THE EARTH IS A GARDEN

ARI NOVY, PETER H. RAVEN, AND HOLLY H. SHIMIZU

The most visible planetary evidence of the Anthropocene is the tremendous extent to which we develop land to suit our various needs. In terms of raw area impacted, our single greatest land use is for growing plants to feed ourselves and our grazing animals, the act of gardening on a worldwide scale. In the residential and aesthetic context, we often call this practice *horticulture* and in the extensive or rural context *agriculture*, although these terms are by no means mutually exclusive or clearly delineated. Regardless of terminology, we are gardening the planet at a historically unprecedented scale to provide benefits to ourselves. Agriculture, which occupies more than half of the land in the United States and over a third of land worldwide, must form a part of the sustainable whole on which our survival depends. The land modifications we make for agriculture and horticulture are primary determinants of our sense of place, and hence identity, as well as our capacity to feed, clothe, fuel, and provide other staples to a rapidly growing population on a hot and hungry planet.

Historically, clearing habitat for agriculture, while necessary for sustaining humans, has been the biggest enemy of biodiversity and the major cause of its loss. Some 12 percent of Earth's surface is devoted to cropland and another 26 percent to grazing. The explosive growth of the human population—from about 1 million 10,000 years ago through 1 billion around 1810 to 7.4 billion today—together with an even more rapid growth in consumption, is challenging the sustainability of Earth's resources. Over the years, we have averted widespread famine by increasing agricultural yield, through the expansion of lands devoted to agriculture, the application of new technologies, and the genetic improvement of crops. Notwithstanding these advances, some 750 million people remain malnourished, with about 100 million on the verge of starvation at any time.

Projections for feeding the world in the future are alarming. The Population Reference Bureau predicts that the world population will increase to

9.9 billion by 2050. The Global Footprint Network estimates that we are currently using about 164 percent of the sustainable productivity of our planet, which makes the future look very challenging. A stable population and sustainable levels of consumption are essential to species success.

Global climate change makes meeting the challenges of the next few decades and beyond even more difficult. Following meaningful global agreements in December 2015, the overall increase in temperatures may be limited to 2°C (3.6°F), though this seems unlikely. But even an increase of just 2°C would negatively impact the world's major agricultural zones. How, then, will we feed more people?

We must determine how much food will be needed and find ways to produce and distribute it. Critically, we must frame this issue with a sense of attainability. The current public discourse on the future of agriculture is heavily dominated by overly simplified or ideological positions that favor specific processed-based solutions, such as preferring organic, urban, hydroponic, precision, or genetically-modified-organism-based agriculture. A more pragmatic approach would circumscribe the desired outcome without favoring or eliminating any specific solution. We might frame the challenge as a set of goals, a proverbial moonshot, such as providing nutritious food for eleven billion people by 2100 on 20 percent less land than is currently used and with decreased inputs, such as fertilizer and pesticides, both per hectare and per unit of production.

Widespread public discussion and agreement will be needed to achieve this agricultural goal for the world. With more than 80 percent of Americans living in cities and more than 60 percent of people in urban environments worldwide, we must educate urbanites about the ecosystems they depend on so they can participate in democratic processes that will lead to successful solutions for the problems of the Anthropocene.

With such a large urban population, the most common contact humans have with plants is through horticulture in city landscapes. Therefore, we must improve our horticultural design and management in urban centers so as to provoke thoughtful and accurate understanding of the often invisible agricultural and other ecosystems that sustain us. In other words, we must garden wisely in the cities so that we will know how to extend that wisdom to the rest of the planet.

Horticulture has traditionally been design driven, with its applied science dimension a means to achieve a desired aesthetic. In the Anthropocene, horticulture must reprioritize so that every landscape is not only pleasing and artistic but also demonstrative and functional as a complete ecosystem.

The traditional concept of beauty in a garden must change. Gardens are often judged by their tidiness, with everything seeming clean and neat. All fallen leaves and branches are removed and each plant is assigned a distinct location. Yet fallen leaves provide cover for roots and habitats for insects and fungi and add to the overall richness of the site while building soil. We must move from expecting gardens to be so tightly controlled and static to requiring that they represent ecosystem processes (e.g., biogeochemical cycling, succession, and evolution). To achieve environmental and horticultural goals, landscape design and architecture must integrate ecological principles with creative garden design so that ecosystem-based gardens can also be aesthetically pleasing. In addition, agriculture must be embedded within urban horticulture in a way that demonstrates how extensive agricultural systems function. Urban agriculture has become an important trend that offers urbanites community cohesion and an exciting introduction to the principles and science of food production, although it enhances food security only in limited circumstances. While urban agriculture will certainly become an enduring feature of cities, we must also make a concerted effort to bring the highly productive agricultural landscapes of the extensive rural lands into the urban tapestry. This more extensive, "agriculture in an urban setting" approach has the potential to present urbanites with realistic scenarios for provisioning all people with food while coexisting in harmony with functional ecosystems and reinforcing the fact that cities are dependent on lands very much out of sight.

Horticulturists and landscape designers must learn how to mimic natural ecosystems and production agriculture in cities. Among other benefits, this will teach citizens what these systems are and how they work. Such urban gardens, by creating the opportunity for education about agriculture and wildlands, foster inclusive and democratic decision making. These practices also increase the environmental integrity of the city itself, by promoting landscapes that reduce pollution, process clean water, support wildlife, and provide beauty and serenity for inhabitants. Ideally, people will no longer think of urban plantings as static design elements but rather as complex, regenerative, and evolving ecosystems. Through designed horticultural mimics, these gardens will open the eyes of the general public to the biological processes governing agricultural and natural ecosystems. We should all realize that our management of both the easily visible landscapes in the cities and the hidden landscapes dominating the rest of the planet holds the key to both our own future and the future of the planet.

As the University of Pennsylvania ecologist Dan Janzen pointed out a half-century ago, "The world is a garden, and we're all its gardeners." Along with achieving a stable population and socially justifiable levels of consumption, we must evolve our gardening to fit sustainably and securely into the global system. If we fail to feed our unprecedented and growing population, people will suffer and die. If we fail to adapt our agriculture and horticulture practices to alleviate the pressures we are putting on the environment, a huge proportion of biodiversity that now exists will be lost forever. The clock is ticking, yet reasons for optimism exist, as evidenced by our continued survival as a species and our increase in food security over the past century. We need to start paying much closer attention to our planted planet; as Voltaire's Candide said, "We must cultivate our garden," though in our case we must also act quickly and wisely.

HUMAN HEALTH IN THE ANTHROPOCENE

GEORGE E. LUBER

As humans entered the Holocene epoch, marked by the end of the last major glacial advance, or ice age, some 11,700 years ago, a planetary transformation began to occur. Increasingly, people lived in permanent settlements, moved away from foraging and hunting for food, and made farming the major mode of subsistence. In the intervening millennia, much of the planet has changed as a result of the quickening pace of human activity and population growth. Today, almost a third of Earth's arable land has been converted to cropland or pasture, more than 90 percent of monitored fisheries are harvested at or above sustainable yield limits, about half of annual available freshwater is appropriated for human use, and species extinction rates are more than a hundred times greater than has been observed in the fossil record.

Perhaps the most startling of all of these global transitions has been the profound alteration of Earth's atmosphere; concentrations of major greenhouse gases—carbon dioxide, methane, and nitrous oxide—are at their highest level in more than eight hundred thousand years, in large part due to the burning of fossil fuels and large-scale land-use changes. The scope and scale of these planetary changes have prompted several in the scientific community to call for the designation of a new epoch, the Anthropocene.

While many of the planetary changes that the Anthropocene represents have consequences for human health, perhaps none will have as broad and transformative an impact as climate change. From weather extremes that directly cause injury or death to potentially profound changes in disease ecology and geography brought about by large-scale state shifts in Earth's ecosystems, climate change will be among the defining issues for public health in the twenty-first century.

Climate change poses a threat to human health in a variety of ways: injury and death from heat waves, extreme storms, and reduced air quality from increasing ozone, aeroallergens, and drought-related wildfire smoke;

illnesses transmitted by food, water, and vectors such as mosquitoes and ticks; and mental health impacts subsequent to disasters. More important, climate change will threaten the critical systems and infrastructure we rely on to keep us safe and healthy: communication and transportation during emergencies, food and water systems during drought, energy grids during prolonged heat waves. As the magnitude and frequency of extreme weather events increase, the resilience of these systems will be tested, and vulnerabilities will be exposed. It is in this sense that climate change will serve as a risk multiplier, amplifying both the exposures that bring about health risks and the vulnerabilities to these exposures.

While climate change will be an important driver of human health, it does not work independently. Other trends that define the Anthropocene—such as urbanization, pollution, and land-use change—will work together with climate change to intensify the threats to human health. Perhaps the most defining feature of the Anthropocene for humans will be the dominance of the urban setting as the primary habitat for large populations. Today, more than half of the planet's population lives in cities; fifty years ago, that number was only 30 percent. This unprecedented and rapid urbanization has led to a landscape transformed from native vegetation to an engineered built environment, resulting in significant differences between the urban climate and adjacent rural regions, a phenomenon known as the urban heat island (UHI) effect.

The UHI can exert a strong influence on local climate. The combined effect of the high thermal mass provided by concrete and blacktop roads, the exaggerated rainwater runoff resulting from the scarcity of permeable surfaces, the low ventilation ability of the urban "canyons" created by tall buildings, and the point-source heat emitted from vehicles and air conditioners magnifies the multitude of impacts brought about by climate change. The health consequences of heightened UHI exposure include rises in both heat-related illness and respiratory illness due to an increase in tropospheric ozone and particulate matter. In a real sense, cities and climate are coevolving in a way that serves to amplify the health hazards resulting from extreme heat, degraded air quality, heavy precipitation runoff, and coastal storms.

In addition to the UHI effect, cities are sources of emissions, as a result of the high density of their housing, motor vehicles, and industry. This combination creates what is called the urban CO_2 dome: higher carbon dioxide levels in cities than in the surrounding rural areas. While climate change is expected to alter the spatial and temporal distribution of several key allergen-producing plant species by lengthening their growing season and shifting their distribution northward, increased atmospheric carbon dioxide concentrations, independent of weather and climate effects, have been shown to stimulate pollen production. A series of studies found an association between elevated urban carbon dioxide concentrations in Baltimore and faster growth and earlier flowering of a ragweed species (*Ambrosia artemisiifolia*), along with greater production of ragweed pollen, leading, in some areas, to a measurable increase in hospital visits for allergic rhinitis. Other studies have shown that poison ivy (*Toxicodendron radicans*), another common allergenic species, responds to high carbon dioxide levels with increased growth and biomass. Its reaction exceeds that of most other woody species and also results in a more potent form of the primary allergenic compound urushiol.

Urban areas also focus risks to their populations from climate change by virtue of where they are located. Cities tend to be found in specific topographic settings: in valleys or basins, at low elevation, and, especially, near coasts. About 40 percent of the U.S. population lives in coastal cities, which are at increasing risk of multiple related threats, including heat waves, flooding from heavy rainfall events, and inundations caused by a combination of sea-level rise, storm surges, and heavy rainfall—all functions of climate change. Their vulnerability increases when the effects of climate change interact with preexisting stressors, such as aging infrastructure, high population density, and poverty. The potential failure of critical components such as water and sewage systems, roads, bridges, communication networks, and power grids increases with climate change and can result in significant threats to health.

Responding to the emergent threats posed by climate change, urbanization, vulnerable infrastructure, and demographic transition—hallmarks of the Anthropocene—requires rethinking the types of actions and policies we adopt to manage the harmful exposures associated with these trends and each population's underlying vulnerability to them. An integrated, health-focused co-benefits approach can provide opportunities to improve environmental conditions while promoting healthy behavior. Its actions and policies can take many forms, including large-scale efforts to decarbonize power generation, "green" urban infrastructure, and transform communities to foster active transportation, and individual-level behavior-change efforts to promote low-carbon, healthy lifestyles.

For example, transitioning to solar, geothermal, or wind power not only reduces greenhouse gas emissions but also improves regional air quality, with benefits to respiratory and cardiovascular health. Policies to promote active transportation, by providing bike paths, walkable communities, and mass transit, not only lower motor vehicle emissions but also promote physical activity, with direct benefits to cardiovascular health and obesity prevention. More-connected, walkable communities can also reduce social isolation, especially among the elderly, which has been shown to be an effective strategy to protect those most vulnerable during heat waves. Individual behaviors such as adopting a plant-based diet can also help reduce the large greenhouse gas footprint of livestock production while reducing the risk of cardiovascular disease and some types of cancers associated with a diet high in red meat. Certainly, many of these policies will have large economic costs, but so does the true burden, often unacknowledged, of carbon emissions on human health and infrastructure. And these policies may be the best way to address the threats posed by the Anthropocene.

The global transformation of our planetary systems, what we are calling the Anthropocene, is increasingly evident and stands to bring substantial challenges to Earth's future. These include climate change, urbanization,

Plate 1.

Georgia Papageorge,
*Maasai Steppe
Ascending—Convective
Displacement,* 1997.
Oil stick and graphite on
canvas with volcanic rock
and cloth; 94 1/8 x 46 1/6 in.
South African artist and
activist Papageorge
has created numerous
artworks, including this
mixed-media piece, as part
of her efforts to document
and arrest the shrinking of
Mount Kilimanjaro's glacier.
Courtesy National Museum
of African Art, Smithsonian
Institution Museum
purchase 98-19-1.
Photograph by
Franko Khoury.

Plate 2. **Fabrice Monteiro,** *Untitled #1,* 2014. Archival digital print.
Monteiro, a Dakar-based artist, created the 2014 series *The Prophecy,*
in which spirits return to teach sustainable practices. Courtesy of
Mariane Ibrahim Gallery, Seattle.

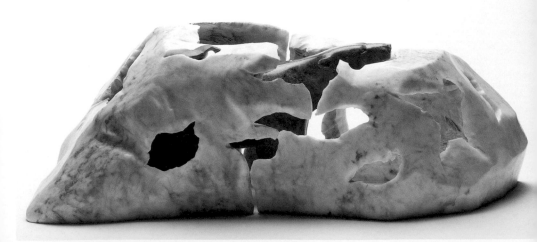

Plate 3. **Fabrice Monteiro,** *Untitled #2,* 2014. Archival digital print. Monteiro's *The Prophecy* series consists of photographs depicting disturbingly beautiful spirits appearing in Dakar's polluted harbors and landfills to educate local populations. Courtesy of Mariane Ibrahim Gallery, Seattle.

Plate 4. **Bill Nasogaluak,** *Arctic Angst* (formerly *Bear Falling through Rotting Ice*), 2006. Stone; 9 x 9 1/16 x 25 9/16 in. Nasogaluak, who hails from Tuktoyaktuk in the Northwest Territories of Arctic Canada, recounts the influence on his art practice of having grown up during a time of great change in Inuit society: "I caught Inuit values in a state of transition, and I can tap all resources—whether from modern society or from tradition." Courtesy Art Gallery of Ontario, Collection of Samuel & Esther Sarick, AGO.87814. © Bill Nasogaluak, 2017.

Plate 5.
Alexis Rockman,
Gowanus, 2013. Oil and
alkyd on wood; 72 x 90 in.
Rockman was inspired to
paint Brooklyn's Gowanus
Canal when a dolphin
swam into the infamously
polluted waterway in
the winter of 2013. His
image reveals the toxic
by-products of urban
development, which will
persist long after the
demise of modern society.
Courtesy of the artist
and Sperone Westwater
Gallery. Photograph by
Adam Reich. Copyright
Alexis Rockman.

Plate 6. **Alexis Rockman,** *Bronx Zoo,* 2013. Oil and alkyd on wood; 84 x 168 in. Rockman explores the theme of human and natural history in his image of New York's historic zoo. The park's neoclassical buildings mimic the ruins of a Greco-Roman city, suggesting a fallen empire reclaimed by the natural forces it once sought to subdue. Courtesy of the artist and Sperone Westwater Gallery. Photograph by Adam Reich. Copyright Alexis Rockman.

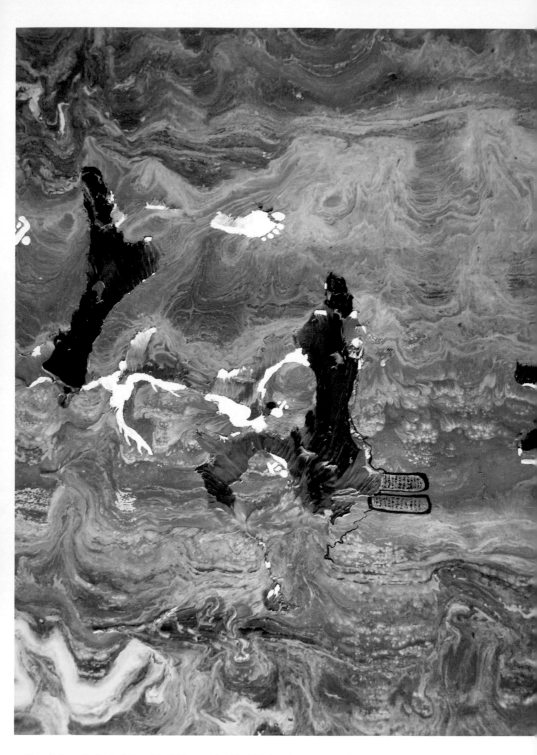

Plate 7. **Jerry Buhari,** *Fall and Spill History* (detail), 2005. Acrylic on canvas, leaves; 60¾ x 51¾ in. Employing autumnal colors and a style that evokes the sheen of an oil slick, the environmentally conscientious Buhari created this work of art to draw attention to the pollution and human rights violations stemming from oil extraction in southeastern Nigeria and their connection to New England homes heated with Nigerian crude. Courtesy of the artist.

difficult demographic trends, and habitat loss and fragmentation. These transformations cannot be looked at as independent phenomena and studied alone but must rather be considered as an intersecting and coevolving set of drivers with great significance for planetary, and human, health.

S teady planetary change lacks the visual drama of singular catastrophic events, such as the types that inspire Hollywood movies
or paintings of volcanic eruptions. The gradual pace and invisibility
of threats such as ocean acidification and biodiversity loss increase
the challenge for artists seeking to communicate about environmental change. Essays in this section describe a variety of approaches.
The paintings of the American artist Alexis Rockman imagine an
urbanized world transformed by extreme climate alterations and
underscore the futility of human efforts to battle nature. In contrast,
contemporary African artists document the continent's ubiquitous
environmental disruptions while campaigning against socially and
ecologically unsustainable choices. In the Arctic, where global warming's early impacts have been the most pronounced and where
indigenous peoples have disproportionately borne the detrimental
effects, depictions of polar bears over the past half century illustrate
how local artistic engagement with climate change has shifted from
an initial celebration of life to mourning and a call for political action.
Visual expression via film and video offers other ways to engage
audiences. Recognizing that people are more likely to support the
conservation of objects and species they have come to care about,
makers of natural history documentaries, for instance, are learning

to balance messages of despair and hope in ways that encourage action and discourage cynicism. In the mass media, where communication of scientific information can be more wholesale than retail, the images and metaphors of visual culture provide essential tools for conveying the complex and often counterintuitive impacts of human activities on Earth's biogeochemical systems. The contributions of artists can stir our collective imagination, capture deep anxieties about degradation of the environment, and fuel the empathy that will be invaluable to attaining a just and sustainable future in an increasingly uncertain world.

✻ THE CITY IN THE SEA
ALEXIS ROCKMAN'S
ANTHROPOCENE IMAGININGS
JOANNA MARSH

Modern megacities are among the most conspicuous signs of the Anthropocene. Towering centers of glass, concrete, and steel have altered the surface of Earth in previously unimaginable ways. One hundred years ago, only 10 percent of the world's population lived in cities, but it is estimated that by 2050, nearly 75 percent of people on the planet will live in a major urban center. This unimpeded growth of the world's cities coincides with climate extremes previously unseen by humanity—heat waves, intense rainfall, and rising sea levels. Cities are at the forefront of global climate change, in both its causes and its effects. Greenhouse gas emissions and energy demand are highest in metropolitan centers, which are also more vulnerable to extreme climate events caused by warming temperatures. In short, the modern megacity has increased the threat of megadisasters. Many experts believe that urbanization and climate change are on a collision course that will either spur new, more sustainable living solutions or exacerbate resource scarcity and endanger already vulnerable communities. As the journalist Gaia Vince writes, "The urban revolution of the Anthropocene could prove to be the solution to many of our environmental and social problems. . . . Or, it could finally prove to be our species' undoing, the apocalyptic version of the dystopian megacity so often portrayed in science fiction."

The artist Alexis Rockman is well versed in these alarming statistics and dire environmental predictions. Over the past twenty-five years, he has created an extensive body of work that combines art, history, science, and popular culture to comment on the myriad ways that humans have altered the planet, from deforestation and biodiversity loss to genetic engineering. Long before Paul Crutzen coined the term *Anthropocene*, Rockman was giving visual form to the transformations—visible and invisible,

108

real and imagined—that have come to characterize the Age of Humans. His vivid paintings imagine a familiar yet fictional world, anchored in disturbing reality.

Among the many locales that he has depicted over the years, New York City holds a special attraction for Rockman. The artist was born and raised there and continues to live and work in the city. He spent his childhood exploring the exhibit halls of the American Museum of Natural History, where he developed a passion for science and natural history. His early work frequently references the displays and dioramas that he saw there, but in recent years the city itself has become a recurring subject. Rockman's first major painting of New York City, *Manifest Destiny* (2004), was also his first work to directly confront the climate crisis and its toll.

Now in the collection of the Smithsonian American Art Museum, *Manifest Destiny* presents a postapocalyptic vision of Brooklyn several hundred years in the future. The borough is inundated by floods or rising sea levels that have rendered the metropolis a semitropical water world. Rockman drew upon extensive scientific research, including consultations with ecologists, climatologists, urban planners, and architects, to create his image of a future New York. This convergence of scientific thought and visionary premonition recalls the style of science-fiction storytelling—a major influence on Rockman's work.

Manifest Destiny sits squarely within the growing corpus of climate disaster literature and film. A subgenre of science fiction, climate fiction (cli-fi) has become a staple of popular culture, evidenced by the release of a veritable deluge of natural disaster films and docudramas that capitalize on our fear of such events and our morbid fascination with their aftermath. The iconography of the contemporary megacity plays a starring role in many of these films, continuing the long tradition of using cities to represent humanity's moral and environmental ills. Rockman applies the same fictive device in his provocative image of a city in the sea.

Manifest Destiny extends twenty-four feet (7.3 meters) in length and is framed at left and right by the ruins of two bridges. On the far left of the composition, a reengineered Manhattan Bridge hovers just below the elevated waters of the East River. Rockman's fictional structure includes a tunnel designed to contend with the city's population boom and changing topography. Mirroring this futuristic artery is the obsolete architecture of the Brooklyn Bridge. Constructed between 1870 and 1883, the Brooklyn Bridge is a potent symbol of the unbridled drive for technological progress that defined the nineteenth century. Rockman also refers to the era's expansionist spirit with the painting's title, borrowing a phrase coined by the New York journalist John L. O'Sullivan in 1845. O'Sullivan used the term as a quasi-religious justification for territorial expansion across the American continent, asserting that the United States had a divine right to conquer and possess westward lands such as Texas, Oregon, and California. The ideology of manifest destiny is deeply rooted in the American sensibility, dating as far back as the Puritan settlers who sought to tame their new frontier and exploit its natural resources. Rockman's painting suggests the persistence of such attitudes today and comments on the deleterious role of global capitalism and urban development in the Anthropocene. It is a damning indictment.

But the end of modern civilization does not mark the end of all life. On the contrary, *Manifest Destiny* teems with organic growth. Rockman uses the image of urban decay to demonstrate the resilience of certain ecosystems and the adaptive powers of nature. Local flora and fauna that have survived the climatic scourge are joined by migrant life from equatorial zones. Rockman's compendium of aquatic creatures raises the question of exactly which species will survive the current biodiversity crisis brought on by humans.

Rockman recently completed two new paintings that explore the interconnected effects that humans have on Earth systems and species. *Bronx Zoo* (2013; plate 6) and *Gowanus* (2013; plate 5) picture two dramatically different sites in New York—one famous for its conservation stewardship and the other infamous for its environmental neglect. Like *Manifest Destiny*, both works imagine a flooded future for Manhattan and its surrounding boroughs. Humans are suspiciously absent, but their imprint is evident everywhere.

In *Bronx Zoo*, Rockman depicts a scene of anarchy amid the ruins of New York's historic zoo. Animals have escaped from their neoclassical enclosures and overtaken the flooded park. Created shortly before Hurricane Sandy hit New York City, it is an uncomfortably prescient image of destruction. The exact cause of the destruction remains unclear, but the flooding suggests a severe climate event. The visual and conceptual parallels with *Manifest Destiny* are undeniable, in particular the allusion to nineteenth-century ideals.

Zoos are a product of the same nineteenth-century impulse that fueled westward expansion across the United States—the desire to conquer and impose order on the natural world. In the era of their inception, however, these constructs were important indicators of the growing acceptance of ecological science and the attendant concern over wildlife depletion. It is no coincidence that the first zoological parks were established in major cities, such as London, Paris, and New York. In these rapidly industrialized areas, "importing" nature satisfied the dual goals of reintroducing wildlife into an urban environment and rescuing endangered species whose habitats were being decimated in other parts of the world. Over the past century, zoos have evolved from places designed for entertainment and dominion over nature to centers geared toward education and species conservation. Centers such as the Smithsonian's Conservation Biology Institute are critical indicators of the Anthropocene, symbolizing the impact that humans have already had on species loss and the role that humans must now play in protecting and preserving biodiversity on the planet.

While *Bronx Zoo* alludes to the dangers of neglecting our environmental responsibilities, Rockman takes this thought experiment even further in his terrifying depiction of New York's Gowanus Canal. Once a thriving tidal estuary nestled in South Brooklyn, the Gowanus Creek was converted into a canal in 1869 to aid transportation and promote industry. Since its completion, serious contamination problems have plagued the canal and surrounding areas. Designated a Superfund site by the U.S. Environmental

Protection Agency in 2010, it is one of the most polluted bodies of water in the United States. A putrid reminder of New York City's industrial past and a cautionary tale for future urban development, *Gowanus* epitomizes Rockman's artistic response to the Anthropocene: dark, toxic, apocalyptic.

The taint of our modern megacities will leave an indelible mark on Earth's geologic record long after they have crumbled into the sea. The question is: how can we forestall the environmental and societal decline that Rockman portends? The answer may reside in the very same cities, where problems of the Anthropocene are most acute. By harnessing their creative and intellectual capital, along with economies of scale that promote efficiency, cities are poised to devise innovative solutions for sustainable living. Artists such as Rockman who work at the intersection of art, science, and imagination will be essential to this process. Rockman's monumental paintings bring the future into the present, helping us grapple with current planetary changes and serving as a catalyst for social and environmental reform.

⚹ AFRICAN ART AND THE ANTHROPOCENE
KAREN E. MILBOURNE

While on a short-term residency in Vermont in 2005, the Nigerian painter Jerry Buhari poured pools of color that reflected the autumnal splendor outside his window onto a stretched canvas and upended it. Sage, gold, burgundy, and brown flowed down the surface of the canvas as the artist moved his brush back and forth with feathery motions to create an effect that evokes the prismatic sheen of an oil spill on the waters of Nigeria's oil-rich Niger delta. With this gesture and these colors, Buhari collapsed the five-thousand-plus miles that separate Vermont from his homeland and the mental barriers that segregate the pollution and human rights violations of southeastern Nigeria from the warmth of New England homes heated with Nigerian crude. *Fall and Spill History* (plate 7), Buhari's visually lush and intellectually charged painting, is emblematic of artworks by artists consciously and conscientiously engaged with living in the Anthropocene—artists who are, to borrow the words of Heather Davis and Etienne Turpin, concerned with "environmental justice thinking, asking what worlds we are intentionally and unintentionally creating, and what worlds we are foreclosing while living in an increasingly diminished present."

Jerry Buhari's commitment to addressing environmental issues with his paintings, prints, installations, performances, and writings is long-standing. To accompany a 2009 solo exhibition, *Man and Earth*, at the Goethe-Institut in Lagos, the artist wrote, "Today the talk of the world is about an endangered Earth. One often wonders how much of the talk is backed with genuine concern and the will to take positive steps. But it should not surprise the world that artists are on the forefront of the discussion of the environment. They have always been." For Buhari, as for other artists around the planet, his artworks are the products of years of research and concerted effort to create sustainable practices and understandings that cross geographic, linguistic, and cultural divides.

Like Buhari, South Africa's Georgia Papageorge feels driven to make works that document environmental change and advocate for new policies and behaviors on an emotional level. In 1996, this passionate activist and grandmother climbed Tanzania's Mount Kilimanjaro ("Mountain of greatness")—the world's tallest freestanding volcanic mountain, which supports a unique, self-contained ecosystem of arid plains, savanna, mountain forest, alpine desert, and glacial ecological zones. When Papageorge returned there nine years later, she was shocked by the rapid and dramatic changes it had undergone. Over the next eleven years, she returned to the mountain numerous times—climbing it four more times and twice hiring biplanes to film its crater from above—and has meticulously tracked down and reviewed photographic and documentary evidence of its condition dating back to 1898. The resulting artworks include an immersive video, dramatic installations, and subtle mixed-media collages, including *Maasai Steppe Ascending—Convective Displacement* (1997; plate 1), in the collection of the Smithsonian's National Museum of African Art, that reveal this mountain to be a barometer of climate change. In *Maasai Steppe*, the artist reproduces images of Kilimanjaro across the decades to illustrate the shrinking of its glacier. Papageorge interrupts these images with vertical striations made with volcanic ash from the mountain, which she refers to as "running tears for an irreparable loss." The vertical lines also evoke a ladder, to suggest the human capacity to transcend or overcome our mistakes.

Both Buhari and Papageorge were included in two high-profile exhibitions that traveled the United States: *Environment and Object: Recent African Art* (2012) and the Smithsonian's *Earth Matters: Land as Material and Metaphor in the Arts of Africa* (2013). *Earth Matters* united more than forty artists from across the vast African continent to explore ways that African individuals and communities have drawn power from, interpreted, and protected Earth. As a result of this exhibition, the artists have joined forces and plan to petition for United Nations Educational, Scientific and Cultural Organization recognition of their efforts. And both *Environment and Object* and *Earth Matters* were preceded by *Rencontres*, the pan-African photography biennial held in Mali in 2011, which took as its theme "For a Sustainable World." Each of these endeavors points to the increasing attention being paid to issues of sustainability by both artists and the art world, and the mounting pressure from African voices to be included in global conversations on the issues defining life in the Anthropocene. Established artists such as Buhari and Papageorge are being joined by a new generation of artists whose visions and approaches suggest the increasing power of the arts to effect change in locations such as Dakar and beyond.

Fabrice Monteiro and Sam Hopkins exemplify this new generation. A former professional model raised in Benin, Monteiro moved to Dakar in 2011, bringing his fashion-world knowledge and aesthetic to the problem of waste management in Senegal. Working with the fashion designer Doulsy (Jah Gal) and the crowdsourcing site Ecofund, the artist released *The Prophecy* (2014), a series of photographs in which disturbingly beautiful spirits, or djinns, emerge from damaged landscapes, as part of an effort to teach sustainable practices to local populations for whom djinns are a guiding force. In *Untitled #1* (plate 2), a majestic figure in a rainbow skirt made of discarded plastic bags, drink cartons, snack wrappers, and other trash towers above the smoking field of rubbish that has transformed the formerly green marshes of Mbeubeuss in Dakar. In *Untitled #2* (plate 3), an aquatic creature with features occluded by black plastic emerges from the once-pristine Hann Bay, where abandoned ships now mix with spilled oil and the blood and offal from a nearby slaughterhouse. In addition to hanging in galleries and museums worldwide, these images will appear in children's books and outreach programs in Senegal. As the artist says, "When it comes to speaking about environmental issues, either you get alarming numbers and statistics or pictures of devastated landscapes. But with projects such as *The Prophecy*, you can speak to the hearts of people by mixing facts and art. Giving this issue a mystical element helps with awareness, and pushes people to change—and change now."

Plastic also features centrally in the work of Kenya's Sam Hopkins. In his 2016 performance and exhibition *The Rubbish Companion*, at Galerie

Börgmann in Cologne, the artist imagined a future in which plastic bags are so rare that there are educational campaigns related to their forms and uses. With savvy humor, the artist compels us to think about what we value, what we save, and to whom we listen. And, importantly, he forces us to think about the future—a period often denied to Africa's artists and citizens when they are, all too frequently, associated with ideas of a romantic past or a troubled present.

Whether bridging cultural distance, recovering valuable historical documentation, speaking to local populations, or envisioning an alternative future, many works by Africa's artists align with what the psychologists Thomas Doherty and Susan Clayton call "a 'beyond toxicity' perspective," in which "the challenges of climate change may also 'galvanize creative ideas and actions in ways that transform and strengthen the resilience of and creativity of community and individuals.'" For as the writer and activist Susan Sontag taught us about images of trauma, it is not the job of these artworks to make us feel nobler for having looked at them or to absolve us of further responsibility; they are a call to action.

❈ WHY POLAR BEARS?
SEEING THE ARCTIC ANEW
SUBHANKAR BANERJEE

In the long list of social-environmental threats that the hotly debated and controversial term *Anthropocene* attempts to hold together, climate change looms large. The Arctic is its bellwether, which is no surprise, as the top of Earth continues to warm at a rate of at least twice the global average. Igor Krupnik and Dyanna Jolly's 2002 edited volume, *The Earth Is Faster Now*, highlights how climate change was already causing significant social-ecological disruptions in the Arctic at the turn of the twenty-first century, even though it was not yet having that troublesome effect on most of the rest of Earth's inhabitants.

Art and popular visual culture can help us apprehend how climate change has transformed the Arctic. In 1973, the acclaimed Inuk artist Pauta Saila made an exuberant stone sculpture, *Dancing Bear*. It depicts a polar bear standing on one of its hind legs, with the other one up and truncated, its front legs slightly stretched to the side and truncated, its face aslant, looking up—a jubilant anthropomorphic portrayal of the animal dancing, with a touch of humor. Almost a quarter century later, on the cover of Thomas Mangelsen's 1997 photo-essay book on polar bears, *Polar Dance*, two bears face each other standing upright on their hind legs, with their front legs stretched to the side—immediately bringing to mind Saila's iconic *Dancing Bear*. Visual depictions of the Arctic produced during the second half of the twentieth century primarily celebrate the Far North and its diversity of life and cultures. But as the new century arrived, visual depictions began to transform—from celebration to mourning and a call to action.

Three years after *Polar Dance* was published, I went to Churchill, in subarctic Canada, where Mangelsen had taken most of the photographs for his book. There I photographed a rather ghastly scene—one bear eating another, behavior not normal for polar bears, the local people said. Because of rapid warming, the sea ice in Hudson Bay is forming later in autumn and melting sooner in spring, which forces the bears to spend more time than usual on land, with a greater risk of starvation. While it is difficult to ascertain a cause such as climate change for one particular incident like the one I witnessed, it is possible that starvation brought about by warming might have contributed to the killing of one polar bear by another. The following year, along the Beaufort Sea coast in northeast Alaska, I photographed a polar bear approaching a whale bone left from the previous year's hunt by the Iñupiat people of Kaktovik. On March 19, 2003, the U.S. senator Barbara Boxer used that photograph to argue successfully against oil drilling in the Arctic National Wildlife Refuge. At the time, the subarctic polar bears in Churchill were suffering from starvation, but the Arctic polar bear population in the Beaufort Sea was thought to be stable.

The following year, however, the scientist Charles Monett and his colleagues observed four drowned bears in the southern Beaufort Sea. In 2006, *Polar Biology* published their research, the first on the plight of polar bears in a warming Arctic sea. The rapid melting of Arctic sea ice was creating vast areas of open water during summer months, which forced the bears to swim much longer distances, in some cases leading to their death by drowning. That year brought wide attention to polar bears: the former U.S. vice president Al Gore and the filmmaker Davis Guggenheim turned the science of polar bear drowning into an animation for their Academy Award–winning documentary, *An Inconvenient Truth*, and an image of a polar bear on thin ice looking tentatively at what lies ahead made the cover of *Time* magazine. Likely aware of the media coverage, the Inuk artist Bill Nasogaluak made an affective stone sculpture, *Arctic Angst* (formerly *Bear Falling through Rotting Ice*; plate 4), that year. Instead of standing on a chunk of sea ice (as on the cover of *Time*) or looking for a piece of sea ice to rest on (as in *An Inconvenient Truth*), Nasogaluak's bear is *inside* the ice; the artist compresses the vast space of the Arctic Ocean into something intimate, reflecting the inextricably linked fates of the bear and the sea ice. Only a small part of the bear is visible through the disintegrating ice, as if the animal were trapped inside a cage from which no escape was in sight. *Arctic Angst* is an extraordinary work of art, in which the entrapment of the bear in its own collapsing home is not unlike the entrapment that many poor and marginalized human communities are already beginning to feel as rapid warming challenges survival and migration is not always a viable option.

While the image of the polar bear was transforming from dancing to drowning, the celebrated Inuk artist Annie Pootoogook created *Bear by the Window* (2004), an unsentimental ink and pencil crayon drawing that challenges popular depictions that romanticize, exotify, or dramatize the animal. She portrays the bear as a scavenger, reaching toward a toppled trash can outside a government-built home, but as it looks up it sees the partially visible face of a child through a crack in a window curtain—the curiosity of the bear meets the curiosity of the child. The drawing depicts the bear as a part of contemporary Inuit life and resists any attempt to turn the animal into an object of spectacle, which happened in Churchill and is now happening in Kaktovik. There is an irony in the rush to see vanishing polar bears, as long-distance Arctic ecotourism contributes to further warming of the Arctic. And such a rush can last only until there are no bears left; in the part of the Beaufort Sea near Kaktovik, their population already declined by 40 percent between 2001 and 2010.

Until recently, the Arctic was thought to be a pristine place, but this is no longer true. Even before polar bear cubs "leave the safety of their dens" in some parts of the high Arctic, the environmental journalist Marla Cone writes, "they harbor more industrial pollutants in their bodies than most other creatures on Earth." Cone calls this tragic transformation "the Arctic Paradox" in her award-winning 2005 book *Silent Snow: The Slow Poisoning of the Arctic*. Through ocean and air currents, industrial pollutants, including persistent organic pollutants such as polychlorinated biphenyl and DDT, from all over the world travel to the Arctic and enter the food web, and then move up the food chain through bioaccumulation, eventually concentrating in the fat tissues of the top predators—polar bears and humans. Arctic warming is speeding the transfer of toxins from ice to water and from there into the food web. The intertwined, tragic fates of the bears and the Inuit underscore the differential nature of accountability and vulnerability in cases of environmental violence and challenge the universalizing tendency of the Anthropocene narrative.

The Iñupiat interdisciplinary artist and activist Allison Akootchook Warden, who hails from Kaktovik, brings us back to the dancing and drowning of polar bears in her solo performance *Calling All Polar Bears* (2011). Its story goes beyond the plight of polar bears, connecting Arctic warming with the relentless push to drill for oil and gas in Arctic land and seas. In this piece, Warden impersonates human and nonhuman, mixes modern and Iñupiat clothing, and sings pop and hip-hop, blending humor with sadness. Her performance is both an elegy and an act of resistance. The Anthropocene, if we begin it at the Industrial Revolution, is largely a product of affluence and power, in which the poor and the marginalized, including indigenous peoples, have contributed little to climate change so far but will continue to be disproportionately affected by its devastating consequences. Indigenous peoples all over the world are fighting back, however, using their art, creation myths, literature, and other stories to resist the destruction of their homelands, food, and culture. *Calling All Polar Bears* belongs to this cosmopolitics of indigenous resistance.

In the climate change documentary *This Changes Everything* (2015), the author and activist Naomi Klein says that images of "desperate polar bears" make her "want to click away" from the films that include them, fueling her search for a new way to tell the story of climate change. Such a reductive assessment fails to capture the complexity in images of polar bears—achieved through visual depiction, literary allusion, memory, and performance—which collectively stand as an emblem of political ecology, connecting the bear and its home, the Arctic, with its people, the Inuit, and the rest of the planet in many unexpected ways, encouraging us all to see the Arctic anew.

❈ THE RETURN OF THE BOOMERANG

LUC JACQUET

My interest in environmental filmmaking began when I was twenty-four. I left France and spent more than a year in Antarctica, where I thought I could get away from every trace of civilization, live at last where all was virgin, unknown, unexplored, intact. This was not to be. Even down there, at the end of Earth, my scientific colleagues were measuring radio-active fallout in freshly fallen snow; we were being irradiated through an ozone hole that gaped over our heads and had been caused by our shaving cream spray cans; traces of pesticides and heavy metals were turning up in the fat of penguins; and glaciologists were drilling into glaciers and telling us about the rise in atmospheric carbon dioxide (CO_2) with ever increasing confidence from evidence in air bubbles imprisoned in the ice.

So I couldn't escape. I was born too late to live on a generous and resilient Earth that permitted our species all excesses, defilements, and predations. I discovered the concept of the raft floating in space, bearing a human race suddenly obliged to be responsible for its acts. I found myself a child of the Anthropocene, an era whose hero is humankind, wretched and miserable. I would have to live with, and probably in opposition to, a society that, to judge by my expertise as a scientist and traveler, was headed for disaster. The system into which I was born and the values that had been handed down to me were, from all evidence, lethal. An uncomfortable finding, to say the least.

And yet how successful humanity has been: after fifty thousand years, this fragile and widely scattered primate has succeeded in a practically complete conquest of the entire planet. All environments, however hostile—hot, cold, dry, on steep slopes—have been colonized or otherwise exploited. Only Antarctica, protected by its belt of unconstrained oceans, has escaped human colonization, but even then, it still has not escaped human influence.

Translated by Roger Sherman.

So, a complete success. Why give up the conqueror's drunken spree after the battle?

In nature, the outcome of this type of success is well known: when a pioneer invasive species, capable of adapting to all situations and driven by a formidable instinct for conquest and a high birthrate, becomes too numerous, it may disappear brutally, starved out or choked by its own waste—the return of the boomerang.

How long before our own boomerang comes back? For us it's more complicated, because we are aware of the phenomenon and have the technical means and the energy to salvage the situation. Still, nothing is being done, except maybe at the margins. One idea obsesses me. We have in our genetic heritage, and thus deep within ourselves, the undoubted qualities of conquerors, discoverers, domesticators. We are also motivated by permanent fear of the future, panics from the abyss of time: Will I eat tomorrow? Will I be able to shelter and feed my family? Survive the calamities and chances of existence? Everything that we have done for millennia tends to make the world more predictable, less dangerous. We love to foresee.

So, will we be able to change this colonizing attitude into an attitude of stewardship? This question is the colossal challenge that faces our species; furthermore, we are not in the habit of making decisions in the name of everyone collectively but rather do so in a fragmented fashion, according to the interests and beliefs of the group, the family, the clan, the nation. A dizzying catalog. How, then, can I situate myself as an individual within this superhuman program?

First, with emotion. In the forty-eight years of my life as a filmmaker, I have seen the world degrading around me, landscapes disappearing. I have seen nature just trickle away. I know that the food I eat is unwholesome; my lungs let me know every day that the air I breathe is polluted; the water I drink resembles a chlorine tea; a blue sky is an indication of a high-pressure area and so of pollution, which the pollution warning signs on the highway are there to remind me of. It's all noxious, even if not immediately visible, and depressing. When I pick up my camera, I often have trouble framing a shot to avoid traces of civilization or scars on the landscape, and I take it hard—because, like everybody else, I need nature to regenerate myself, to dream, to tell stories. I need to feel at peace with the world in which I live without bearing the collective curse of the destroyer. Of course, in saying this I acknowledge that we must first ensure that our more basic biological needs such as nourishment and breathing are safeguarded.

Next, I do my share by fulfilling my responsibility to others. I have three children; I owe it to them to hand over a viable and intact world. I want to share with them places that are dear to me. Besides, I feel that I have had a lot of good luck in my life, and consequently that I have a responsibility to share this luck. I don't like politics; my friends often call me a bear that prefers the retreats of its mountains to showing itself, and they are right. Yet I feel that I don't have a choice, as my grandparents did not when the Nazis invaded their country. Submit or resist?

It's almost like fighting myself. I come from a peasant tradition where nothing is more satisfying than land subjected through full struggle to the will of the one who cultivates it. I learned to be proud of our victories against "weeds," against "harmful" species, against wasteland and sloppiness, proud of a land that was "productive," of a forest finally "clear-cut," of a swamp "reclaimed," and so on. We also learned that technological progress was king and science sovereign, that these two divinities conferred on us colossal—maybe divine?—powers. And here we are, betrayed by these same gods, which are turning against us.

Who today wants to renounce these powers, this comfort, even though they threaten us? Nobody. Here is the immense paradox of ecological discourse, which no one wants to hear but which nevertheless speaks in the name of the well-being of the human race. One would have liked to profit from the victory, enjoy the fruits of the planetary conquest, before going further.

I sometimes feel overwhelmed by the immensity of the effort needed to tip society toward an awareness of a sustainable world. I sometimes think that I am given enough means to wear myself out but not enough to be effective. The inescapable lesson of history is that this is a long-haul effort. It was not those who first took to heart the iniquity of slavery or the injustices done to women and men who had the joy of seeing laws passed that remedied those evils.

We have started a marathon, but we are falling behind. If we slacken our pace, we will lose altogether. Yet I believe in the power of the positive force. We must make people want to help rather than judge them. We must explain, make them understand. Art in general, and film in particular, is a formidable tool for this battle: it speaks the language of empathy and emotion, which is what we need to give ourselves the energy to invent a new form of existence on this planet.

❧ FILMMAKING IN THE ANTHROPOCENE

JOHN GRABOWSKA

My home in North America's Eastern Woodlands rests about ten miles (sixteen kilometers) upriver from where the Potomac joins the Shenandoah and bursts through a gap in the Blue Ridge Mountains to continue down to Great Falls, the Chesapeake Bay, and eventually the Atlantic Ocean. One of my favorite pastimes in late summer is to walk through the woods and climb down the limestone cliffs to fly-fish in the big river for feisty smallmouth bass. Ospreys and eagles soar overhead pursuing the same prey. Bank beavers make an occasional appearance, surfacing, eyeing me, and submerging. Crafty little green herons slink along the rocky bank, where tracks of raccoons, opossums, and (sometimes) black bear are imprinted in the mud.

Hiking through mature woodlands and fishing in the Potomac allow me to indulge in a primordial fantasy. As I stand waist deep in water, casting for smallmouth, seeing not another soul, I can imagine that this stretch of the river is a two-tone wilderness much as it was five hundred years ago, with heavily forested riverbanks, the green river gliding by, the blue sky above.

The fantasy, of course, is just that. The mature trees make up a secondary forest and are known as dog hair woods, growing up crowded and branchless, desperately competing for sunlight after the cattle were removed from what was once pasture. The diversity of plants is diminished because of the absence of fire, the eradication of apex predators, and the resulting proliferation of white-tailed deer, plus the introduction of invasive plants; multiflora rose, barberry, and wineberry catch on my waders as I walk through the woods on the way to the cliffs. The smallmouth I catch and release are nonnative and now intersex, with most of the males growing ovaries because of endocrine disruptors in the water. The sound of jets on the glide path to Dulles Airport is omnipresent: the Lower 48 is a noisy environment, particularly to a filmmaker attuned to such things.

Such contrasts between idyllic visions and the reality of human impacts on the environment have been reflected in the evolution of natural history filmmaking over the past century.

Often considered the first documentary film, Robert Flaherty's *Nanook of the North* (1922) embraces the romanticism of humankind living in harmony with nature. Like many natural history films to follow, it was shot in a remote location, features exotic cultures and charismatic wildlife, and encourages audiences to engage in armchair adventuring. In one of the next major milestones in documentary film, attention to the environment took the opposite approach. Pare Lorentz's *The Plow That Broke the Plains* (1936), a jeremiad against abuse of the land, introduced many techniques that make nature documentaries so popular: liberal use of music to guide emotions; spare, poetic narration; striking cinematography; and deft editing. Some seventy years before the term was popularized, *The Plow That Broke the Plains* was unmistakably about the Anthropocene, profiling the ecological crisis caused by unsustainable agricultural practices paired with Dust Bowl drought.

After World War II, Walt Disney's wildly popular *True-Life Adventures* series, shown in theaters from 1948 through 1960, offered nature as pure entertainment. That tradition of producing armchair nature adventures continued into the television era with the work of the British Broadcasting Corporation's Natural History Unit, the productions of the National Geographic Society, and the Public Broadcasting Service's award-winning *Nature* series. In the United States, the proliferation of cable channels dedicated to wildlife and nature combined with the tyranny of television ratings and the splintered media landscape to increase the emphasis on flash and shock (e.g., the Discovery Channel's *Shark Week*). Before the plethora of cable programs appeared, the prevailing aesthetic in most nature documentaries in the previous hundred years had been a fascination with charismatic wildlife and seemingly pristine landscapes and an appreciation of the wonders and complexity of science and the natural world. That emphasis on beauty and wonder was and is legitimate and valuable, a method of inspiring a love and regard for the natural world. People tend to protect and preserve that which they love, but most of these films avoided showing human intrusion upon the landscape: the telephone wires just outside the frame, the highway in the distance, or the degraded farm field just off camera.

With the advent of the modern environmental movement and civic engagement in the 1960s, protest films revealing environmental crises began to appear, highlighting the disasters of pollution and oil spills, though less frequently pointing out the devastating habitat loss caused by rampant development. Again, before the term came into being, filmmakers were producing documentaries about the Anthropocene without even knowing it.

Crises are a heady brew, however. If audiences are fed a relentless diet of environmental calamity and catastrophe, they shuffle out of the theater dispirited rather than engaged, hopeless and fatalistic in the face of problems too big to solve or even comprehend, or they simply ignore yet another outrage du jour. I have cofounded two environmental film festivals; if their slates are one helping of despair after another, the theaters are empty. The audience votes with their feet, and heads to a pub.

Conversely, whistling past the graveyard is an exercise in irresponsibility. Even the venerated BBC and its partner Discovery Communications were rightly criticized for avoiding any mention of the causes of climate change in their 2012 series *Frozen Planet*. Eliding the issues that threaten the natural world in the Anthropocene is unacceptable.

What is called for is a delicate balance between hope and despair. In my own natural history films, I do not ignore the threats, particularly the single issue of greatest import in the history of humankind, anthropogenic climate change, but I try to balance the grim acknowledgment of current and future realities with an inspirational reminder of what it is we so love about the natural world. In that spirit, I have made films that examine the mid-Atlantic's Outer Banks barrier islands, collapsing because of rampant development and rapid sea-level rise; about a mountain range in the desert Southwest where entire forests of piñon pine have expired because of heat and drought; and about the remote Alaska Peninsula, where bears digging for razor clams on mud flats still unearth lingering oil from the *Exxon Valdez*, twenty-five years after and 450 miles away from the 1989 spill.

Audiences must not be shielded from the realities of the Anthropocene, but they need hope and desperately want to be reminded of what it is about the natural world that they loved as children and want to love today. And there is reason for hope. Al Gore is now making the case for optimism on climate change because of rapid improvements in renewable-energy use and technology. In 2016, Jane Lubchenco gave an address titled "Enough with the Doom and Gloom!" to the National Academy of Sciences. E. O. Wilson's Half-Earth Project is the very definition of audacity of hope. Natural history films must illumine the realities of living in the Anthropocene while reminding audiences that the natural world is precious and valuable, both for how it provides for the very existence of human society and for its own wondrous complexity and capacity for inspiration.

• • •

In early spring, the water of the Potomac is too cold and turbid for fly-fishing, but I walk in the gray and brown woods to see the annual phenomenon of spring ephemerals. Up through the leaf duff, spring beauties come peeping, awakening from the ever shorter winter, followed by delicate cutleaf toothwort and silly Dutchman's breeches. Twinleaf and bloodroot appear and disappear, sometimes in the same day, petals detached by a mere breath of wind.

Despite the encroachment of Japanese stiltgrass and garlic mustard, these native wildflowers return to provide delight. Pileated woodpeckers make their looping, parabolic flights to dead trees; tropical migrants bring their song back to the still-temperate forest; blue jays mock other avifauna with false red-tailed hawk screams. Under a log, a little red-backed salamander hides—delicate, resilient, beautiful. Even in my sad, sorry, lovely woods, the natural world has the capacity to thrill, inspire, and enlighten, and that is what filmmaking in the Anthropocene must do.

❊ PICTURING PLANETARY PERIL
VISUAL MEDIA AND THE ENVIRONMENTAL CRISIS
FINIS DUNAWAY

At a key moment in the surprisingly popular, Academy Award–winning documentary *An Inconvenient Truth* (2006), Al Gore stands in front of an enormous graph. A jagged red line tracks the changing levels of atmospheric carbon dioxide over the past 650,000 years. Then a jagged blue line plots the average temperature readings over the same period. Viewers immediately notice the uncanny similarity between the two lines: every rise or dip in the red seems to generate a corresponding rise or dip in the blue. "I can't think of another movie in which the display of a graph elicited gasps of horror," the *New York Times* film critic A. O. Scott observed, "but when the red lines showing the increasing rates of carbon-dioxide emissions and the corresponding rise in temperatures come on screen, the effect is jolting and chilling." After the lines appear, Gore steps onto a mechanical cherry picker that elevates him to the current carbon dioxide level. The red line moves relentlessly upward: its exponential verticality forecasts a radically changed planet. *An Inconvenient Truth* elicits shock and fear at the capacity of industrial society to transform the climate. The film presents science as prophecy, a warning about the long-term, planetary dimensions of climate change.

More clearly than any other cultural text had done to date, *An Inconvenient Truth* explained to mass audiences how the carbon cycle invaded the news cycle. Throughout the history of popular environmental images, the media have tended to emphasize the sudden violence of oil spills and other spectacular examples of ecological devastation. Climate change—like other problems associated with the Anthropocene—is a fundamentally different kind of threat: not an immediate, highly visible catastrophe but a gradually escalating, often invisible form of environmental danger. *An Inconvenient Truth* and other recent images of planetary peril suggest broader questions about visual culture and the Anthropocene: How can we learn to see systemic problems that lack the obvious visibility of an oil spill or natural disaster? To what extent does the emphasis on picturing planetary change obscure the uneven and profoundly inequitable experiences of the Anthropocene? Does the repeated focus on exponential data curves encourage environmental action or provoke fatalistic views of the future?

The visual media have helped make climate change—a systemic, slow-motion disaster—seem less abstract. In a prime example, the photographer James Balog's Extreme Ice Survey employs time-lapse photography to record the stunning retreat of glaciers around the world; his project is profiled in the award-winning documentary *Chasing Ice* (2012), which includes footage of the largest glacier calving ever filmed. Likewise, *An Inconvenient Truth* borrows from scientific projects that use repeat photography to document glacial recession. These projects pair historic photographs of glaciers with contemporary photographs taken in the same spots: the then-and-now images reveal massive melting over relatively brief periods. Such images spectacularize the unspectacular, placing climate change in historical perspective to dramatize the rapidly changing world. These pictures also reject the news media's emphasis on sudden catastrophe to grapple with long-term, accretive crises. In *An Inconvenient Truth*, Gore shows images of recent weather events from heat waves to hurricanes to demonstrate how increasing temperatures have produced recurring patterns of cataclysmic harm. In another viewing context, such as news media coverage of disasters, the images might be seen as depicting freakish events of nature, unrelated to accumulating carbon emissions. Here, though, they appear as the calamitous result of the graph's ascending red line and resemble, as Gore puts it, "a nature walk through the book of Revelations."

As Gore's innovative use of graphs demonstrates, data visualization provides a powerful tool to communicate evidence of anthropogenic change, yet these displays often suffer from a reductive view of history and a failure to attend to power relations. Consider the Great Acceleration graphs featured in the International Geosphere-Biosphere Programme's influential *Global Change and the Earth System* (2004), by Will Steffen and his cowriters. All of them imagine humanity as a singular, collective entity, marching together as an upward-sloping line that represents massive changes to planetary systems. Whether depicting the exploitation of global fisheries, the frequency of great floods, or the concentration of carbon dioxide and nitrogen dioxide in the atmosphere, these graphs produce a geometric redundancy: in all cases, the data sets rise exponentially after 1950. If they were displayed on a big screen, Gore would need to ride his cherry picker to reach their peaks. Like Gore's climate change graph, the Great Acceleration charts present a linear, almost teleological narrative of *Homo sapiens* as an increasingly powerful but also increasingly destructive force. Seen in this manner, the Anthropocene concept renders the species as a set of data points plotted across shared coordinates of time and space. Anthropocene graphs provide startlingly clear representations of the scale and rapidity of global ecological change. Yet they neither tell us about the causes of our current environmental predicaments nor register the unequal experiences of environmental risk.

In this way, Anthropocene imagery reproduces some of the same problems that have limited the imagination of popular environmentalism: too often, mainstream depictions of the movement have emphasized notions of universal vulnerability and universal responsibility, framing all people, no matter where they live, no matter their class or race, as equally susceptible to environmental harm *and* equally culpable of causing the environmental crisis. From Cold War concerns about radioactive fallout to contemporary anxieties about global warming, the visual media have repeatedly portrayed all people as inhabiting a shared geography of environmental danger. In *An Inconvenient Truth* and other popular environmental works, pictures of the whole Earth—especially photographs taken by NASA astronauts in outer space—signify the planetary scale of the crisis and act as emblems of universal vulnerability. Like the Great Acceleration graphs, though, this planetary perspective obscures the realities of environmental injustice and deflects attention from the power relations that determine ecological inequalities. Moreover, the popular focus on universal responsibility has moved environmentalism from the political to the personal, prescribing individual actions and green consumerism as short-term solutions to long-term environmental problems.

Even as it places climate change within an extensive temporal vision, *An Inconvenient Truth* follows this familiar pattern. "I don't know about you," the popular food writer Michael Pollan commented, "but for me the most upsetting moment in *An Inconvenient Truth* came long after Al Gore scared the hell out of me, constructing an utterly convincing case that the very survival of life on earth as we know it is threatened by climate change. No, the really dark moment came during the closing credits, when we are asked to . . . change our light bulbs. That's when it got really depressing." According to Pollan, the "immense disproportion between the magnitude of the problem" and the "puniness" of Gore's proposed solutions—using energy-efficient light bulbs, carrying reusable bags to the grocery store, and, if you can afford to, buying hybrid vehicles—"was enough to sink your heart." While Scott praised its terrifying graphs, Pollan faulted the film for its failure to fashion a compelling, inspiring vision of the future.

Confronting the crises of the Anthropocene will require more than fantasies of personal empowerment. While popular environmental images have often promulgated green consumerism and personal responsibility, some activists and image makers have sought to move beyond this individualist frame by imagining collective responses to global warming and other environmental crises-in-the-making. Climate activist groups such as 350.org join Al Gore and other Anthropocene theorists in relying on data visualization. Like *An Inconvenient Truth*, 350.org videos use red lines to depict the exponential data curves of rising concentrations of carbon dioxide and, as indicated by the group's name, to warn of the extreme dangers signaled by exceeding the 350 parts per million threshold. Yet 350.org and similar climate activist groups reject the limited model of citizenship embraced by the mainstream media. In the words of 350.org's operations director, Jeremy Osborn, they emphasize that "it's not light bulbs, not Priuses" but "large systemic change" that is truly necessary to reduce greenhouse gas emissions. Climate activists harness visual and social media to galvanize public concern and to question the structure of dominant energy systems, especially the power of the fossil fuel industry. These groups are also working with indigenous peoples and environmental justice activists throughout the world to bring attention to the vast inequities of climate change and to question the universalizing message of whole-Earth imagery. Rather than naturalizing the red lines of apocalyptic despair, rather than succumbing to a fatalistic outlook on the future, these activists are challenging the short-term, profit-making interests of corporations and trying to envision long-term, sustainable ways to live in the Anthropocene.

With recognition of the ubiquitous impact of human-induced environmental change comes the question "Now what do we do?" Some want to run and hide. Some want to continue to ignore what is happening. Some want to celebrate because humans are finally in charge of the planet. Some believe that the natural world will take care of itself once humans have done themselves in. Some want to sit back and let technology provide the solutions. Some do not know what to do. And some are looking for sound solutions that will ensure a better future for Earth and humankind. The answer is neither simple nor straightforward, given the multitude and complexity of the variables at play. Fortunately, scientific and humanistic understandings of the Anthropocene are steadily advancing, providing knowledge essential for devising global efforts to slow the rate of change of biophysical transformations, mitigate their severity, and forge socially just and sustainable adaptations. Maintaining a livable planet with more people occupying a finite space subjected to accelerating environmental stresses will require extraordinary collaborations among citizens, governments, social and religious institutions, the marketplace, and the private sector. The link between the global environmental exigency and the international crisis of inequality is a fundamental aspect of the Anthropocene. As human activities in all societies become ever more enmeshed with landscapes, water-scapes, and the totality of life on Earth, we must learn to better manage our environments and ourselves. Surviving the challenges of the Age of Humans—exacerbated by expanding populations, overuse of resources, and environmental degradation—demands not only increasing knowledge of the natural world but also slowing the rate of species extinction. Much needs to be done, to be done with urgency, and to be done with a spirit of cooperation and sensitivity to our shared future on this Earth.

❧ DRAGONS IN THE GREENHOUSE
THE VALUE OF KNOWLEDGE AND THE DANGER OF UNCERTAINTY

RICHARD B. ALLEY

Disturbing headlines during 2014 warned of unstoppable sea-level rise from collapsing ice in West Antarctica. That ink had barely dried before a distinguished panel of retired high-ranking military officers noted that "the projected impacts of climate change . . . will serve as catalysts for instability and conflict." As a climate scientist reading those headlines, I was reminded of "Here be dragons," written in Latin on the Hunt-Lenox Globe from circa 1510.

Climate science may at first seem like an odd path to unknown but dangerous beasts. Despite occasional public statements to the contrary, the basics of climate and energy are not especially controversial—we gain much good from burning fossil fuel, but the carbon dioxide (CO_2) thus released turns up Earth's thermostat, and the impacts from each degree of warming will cost more than those of the previous degree, so an economically efficient response would start now to take measured action to reduce the rise of carbon dioxide and to adapt to the damages that are still expected to occur.

Uncertainties do encircle this basic consensus, however, just as they have always surrounded any growing body of knowledge. Christopher Columbus's voyages forced mapmakers to confront what they didn't know, too. The Hunt-Lenox Globe, perhaps the second-oldest surviving globe from after Columbus and now at the New York Public Library, has a recognizably accurate South America to go with the well-known classical world, but a few scattered islands stand in for all of North America. And on a vaguely recognizable coast of Asia are the famous words "HC SVNT DRACONES."

Whether the mapmaker was thinking of scaly fire-breathers, Komodo monitor lizards, or something else is open to debate. But the words have reminded us over centuries that new knowledge also highlights what we don't know—early mapmakers couldn't guarantee freedom from dragons in largely unexplored territory. Further research usually cuts off the long tail of possibilities, finding familiar, dragon-free conditions. Occasionally, though, a long-tailed "dragon" may show up.

As the science of climate change has matured, it has shoved many of the possible dragons that we worried about a decade or two ago off the map. The National Research Council reported in 2013 that methane freed from clathrate ice in the sea floor by warming from our burning of fossil fuels is likely to amplify that warming, but slowly—giant methane belches suddenly cooking the planet are very unlikely. Changes in North Atlantic circulation will probably influence our future in unpleasant ways, too, but the flash freeze in the 2004 movie *The Day after Tomorrow* was science fiction, with emphasis on the fiction.

But as new research has lowered these concerns, just as new explorations half a millennium ago replaced islands and dragons with North America on maps, other tipping points in the warming Anthropocene have become more worrisome. Hurricane Sandy brought a storm surge to a piece of American coastline for a few hours in 2012, filling subways in New York and crippling the city; the West Antarctic studies point to the likelihood that changes already have been or soon could be initiated that will commit the world's coasts to long-term sea-level rise similar to or larger than Sandy's surge. This result may still be revised as the modeling is repeated and extended, and so far the onset of any rapid rise seems to be decades or more in the future, but the consequences still loom large to many observers. And as scientists look at the effects of changing climates on ocean acidity and dead zones, and on ecosystems including the Amazon rain forest and widespread coral reefs, many damaging tipping points remain possible.

Furthermore, despite their strength and vitality, our economies include large vulnerabilities that climate change can make worse. For example, in their report, the retired military officers traced the dashed line leading from drought in Syria to civil war there, and from heat and drought in Russia and China through crop failures to rising bread prices and then uprisings in North Africa. The report's comment on conflict in Mali applies to all of these situations and more: "While climate change alone did not cause the conflict, it certainly added environmental stressors."

Solid scholarship shows that in a world without dragons, in which we prepare wisely for the well-expected, smoothly rising impacts of warming, it will be economically beneficial to start now to enact efficient policies that reduce the release of fossil-fuel carbon dioxide into the air. But in a world where a threshold is occasionally crossed, unexpectedly and rapidly triggering the loss of an ice sheet or a huge ecosystem or much of the economy of a trading partner, the costs of warming are likely to be substantially higher, increasing the value of slowing the release of the fossil-fuel carbon dioxide that pushes these things toward their tipping points.

Early mariners had to successfully manage uncertainty to bring back the information for the Hunt-Lenox Globe. Fog was a reason to slow down: no prudent pilot wanted to discover an unknown and unseen reef by sailing into it at full speed. Today one occasionally hears the argument that society should not take actions to slow greenhouse gas warming until we have absolute scientific certainty that we face damaging hazards, that we should sail full speed ahead through the fog of our remaining ignorance, an idea that might have seemed strange indeed to those sailors of yore.

It may be a little self-serving for me as a researcher to point out the value of doing more research, but knowledge really is power and money in this case. Ship captains eagerly embraced radar and sonar, GPS, and better charts as they became available, greatly reducing uncertainty and improving the ability to speed through bad weather. By better predicting what dragons do and don't await us in a warming future, we may be able to prepare for or avoid the most dangerous ones, sailing ahead more rapidly and confidently.

Yet even with all the modern navigational technologies, sailors still carry life-preserving equipment and emergency beacons on ships built to minimize the risk of sinking if a collision does occur, and they carry insurance against disasters. Uncertainty can never be removed entirely, and the slight chance of a devastating accident is sufficient reason to take serious precautions. Science will never learn exactly where all the possible dragons are in a warming future. Analogy to the precautions taken by sailors suggests the value of actions to increase societal resilience to unexpected shocks and perhaps to work harder to slow the warming, just in case.

Exploration since Columbus has accurately filled in most of the geographical unknowns, so the Hunt-Lenox dragons are gone from our globe. We can send at least most of the climate change dragons to join them, and prepare wisely for any that remain, by designing and implementing wise policies using the knowledge we already have and new knowledge we will obtain from targeted research. The sooner we get started, the more effective we will be.

⚐ WHY SCIENTISTS AND ENGINEERS MUST WORK TOGETHER
G. WAYNE CLOUGH

Climate change was an abstract concept for me until I chaired the National Research Council's Committee on New Orleans Regional Hurricane Protection Projects following Hurricane Katrina in 2005. The committee was charged with reviewing plans proposed by the U.S. Army Corps of Engineers for a new system with a price tag of more than fifteen billion dollars. During our discussion of the design elevation for the levees and floodwalls, I asked what the system's life expectation was. After some give-and-take, we agreed it should last at least one hundred years. Because sea levels are predicted to rise two to three feet over that time, climate change instantly moved from a scientific concept to a very real design issue for engineers. And in 2012, Hurricane Sandy demonstrated once and for all that this problem extends to all of our coastal cities, not just New Orleans.

In the ensuing years, awareness of the effects and magnitude of climate change has grown. In 2014, all U.S. federal agencies were required to produce adaptation plans to address the potential effects of climate change. Concerns surfaced about food and water supplies, vulnerability of defense bases in coastal areas, and the likelihood of increased global conflicts. The Smithsonian Institution cited the growing potential for flooding of some of its largest museums in Washington, DC, as a result of increasing levels of storm surges. The cumulative impact of the adaptation reports was a wake-up call. In 2015, the historic Paris Agreement was signed by 195 nations committing to changes aimed at limiting warming to no more, and preferably less, than 2°C (3.6°F) above the average temperature of the preindustrial age.

All of the expressions of concern were important, but open questions remain as to exactly how the driving forces behind climate change can be slowed or reversed, how rapidly this can be achieved, and how this might be done while sustaining a positive world economy. I contend that the answers can be found only through research and the application of knowledge at the intersection of the sciences and engineering.

After retiring from the Smithsonian, where I had been the director since 2008, and returning to the Georgia Institute of Technology as a faculty member in 2014, I made it a point to meet with faculty working on issues relating to climate change. The issues being addressed ranged from national policy to the science of changes taking place in our oceans. I was pleased to learn about new approaches that were being developed to take carbon dioxide from the atmosphere and sequester it below ground. I was impressed to hear of the work of the Carbon-Neutral Energy Solutions Laboratory and the University Center of Excellence in Photovoltaic Research and Education, funded by the U.S. Department of Energy, in developing energy sources that are both carbon neutral and economic. For example, advances in photovoltaic technology reduced the cost of solar energy by 70 percent from 2009 to 2016. Solar costs are now competitive with those of coal and will soon be lower than those of natural gas. In the past several years, well over 30 percent of the new energy produced in the world has come from solar sources.

My engineering friends were rightfully proud of their progress in creating beneficial technologies, but it was apparent they were not aware of a context that would help them appreciate how much difference their work could make in the effort to mitigate the effects of climate change. Would it be enough to restrain the amount of carbon dioxide in our atmosphere, slow the rise of sea levels, and minimize the acidification of our oceans so that our world could sustain life as we know it? This was a question with an answer that lay in the science underlying climate change.

While at the Smithsonian, I had made it a point to spend time with as many of the five hundred or so scientists who work there as possible, often traveling around the globe to observe their work. I witnessed the effects of the oceans' warming and acidification on coral reefs, observed evidence of the threat posed by melting Arctic tundra to the way of life of Alaska's Yupik people, and walked through tropical and temperate forests where the growth patterns of many different species are changing in almost real time.

Every story was compelling, demonstrating how climate change is dramatically impacting natural systems. However, I found that scientists, like engineers, have a blind spot. Few of them were aware of the work of engineers on new carbon-neutral energy sources or atmospheric carbon removal, which could mitigate the effects of climate change. And here as well no one was asking, "What would it take to really make the difference needed?"

If the communications gap between engineers and scientists were to be closed, we could develop a more cohesive and robust strategy to address climate change. At a minimum, a coming together would make both groups better informed and more effective collective advocates for addressing the challenges at hand. The merging of forces would allow both sides to better appreciate the practical issues involved and how economic factors play into the equation.

Steps are being taken to address this problem. At the Smithsonian, an institution-wide initiative led to the development of Living in the Anthropocene, a project offering people from different disciplines the opportunity for dialogue about common issues. In conjunction with the 2015 Paris meeting, two additional major initiatives were announced. First, twenty nations, together accounting for upward of 75 percent of today's carbon emissions, announced Mission Innovation, a promise to invest billions of dollars in public-sector monies to accelerate global clean-energy innovation. As part of America's commitment, the U.S. Department of Energy launched the Clean Energy Investment Center, to make information about energy and climate science from government agencies accessible and understandable to the public. Second, a group of philanthropists and corporations banded together to create the Breakthrough Energy Coalition, to provide a fund of twenty billion dollars to match the public-sector monies of Mission Innovation.

The timing could not be better for scientists and engineers to work together to bring clarity to issues of joint interest and paramount concern to the world. In 1853, the Smithsonian's first secretary, Joseph Henry, said it this way: "James Smithson was well aware that knowledge should not be viewed as existing in isolated parts, but as a whole, each portion of which throws light on all the others, and that the tendency of all is to improve the human mind."

✻ HAZARDS TO OUR HERITAGE
CHOICES AND SOLUTIONS
CORINE WEGENER

Cultural heritage is indivisible from what it means to be human. It is what we receive from our ancestors, what we decide to preserve (or not to preserve) in the present, and what we intend to pass to future generations. It embodies a society's identity and hope. Heritage may be mostly of local significance, such as a war memorial inscribed with the names of local veterans, or it may be internationally renowned, such as Leonardo da Vinci's *Mona Lisa*. All cultural heritage is an expression of our memory, creativity, and innovation, and it has value. But heritage, like Earth itself, is vulnerable to the impact of our activities and the choices we make. The hazards that humans impose may be characterized by three *E*s: environment, encroachment, and escape. The actions that humans undertake to prevent the loss of heritage are represented by three *M*s: mitigation, movement, and memory.

• • •

Hazards to cultural heritage make for a very long list: fire, armed conflict, extreme weather events, rising sea levels, and urbanization, to name only a few. In the past, we typically labeled hazards "natural" or "human-made." These terms are misleading. While humans do not control many of the hazards, we do control how we plan for them and reduce their impact. On the other hand, anthropogenic activity exacerbates each of the three *E*s.

Environment. Although no region in the world is hazard free, certain locations are exposed to increased risk. Much of the history of human activity has occurred near water. Many of the 720 United Nations Educational, Scientific and Cultural Organization (UNESCO) World Heritage Sites are near coastlines, making them vulnerable to coastal erosion, tropical cyclones, and flooding. Climate change is a slow-onset disaster, producing rising sea levels and storms of increasing duration and intensity. Some studies predict that sea levels may rise by as much as six feet (nearly two meters) by 2100, potentially swamping sites such as the city of Venice and submerging whole island cultures. Our environment can also demonstrate its destructive power very suddenly. Earthquakes and volcanoes threaten heritage sites around the world, particularly within the Pacific Ocean's famed Ring of Fire.

Encroachment. Human encroachment is a major cause of heritage destruction. Overpopulation, urban sprawl, and economic development often outstrip a community's ability or desire to preserve historic structures. Legal protections may exist but are often not enforced because of corruption or lack of capacity. Cairo's urban expansion now nearly surrounds the Giza pyramids and the Sphinx, while one of Frank Lloyd Wright's greatest architectural achievements, the Francis Little House on Lake Minnetonka, Minnesota, was dismantled for museum period rooms in 1972 and the rest razed because no one was willing to buy and restore it.

Today cultural heritage is facing a more violent form of encroachment: armed conflict. While the loss of life and human suffering are an incalculable tragedy, the cost to our heritage is also immeasurable. In the wake of the 2012 conflict in Mali, Islamic extremists forbade cultural expressions such as dance, music, and traditional dress, destroyed Sufi religious sites, and burned thousands of the legendary manuscripts of Timbuktu. In Syria and Iraq, conflict and the rise of the Islamic State of Iraq and the Levant (ISIL) have taken a devastating toll. Intense combat has destroyed the Old City of Aleppo, a UNESCO World Heritage Site, and many other sites like it. ISIL has carried out a systematic campaign of deliberate destruction of cultural heritage, smashing collections at the Mosul Museum and razing ancient sites such as Nimrud, Palmyra, and Nineveh. ISIL blew up the Tomb of Jonah (Nebi Junis) in 2014 and then hauled away the rubble, removing all traces of its existence.

Escape. Conflicts in the Middle East and North Africa have spurred a refugee crisis. In Syria, long-term drought, likely due to climate change, was a contributing factor. Drought-stressed rural populations surged into urban centers, exacerbating the economic and political instability that ultimately fueled a civil war. That situation proved fertile ground for ISIL. As often happens in war, local populations resorted to looting archaeological sites to feed their families and help fund their escape, causing irreparable damage. But sometimes ancient sites can provide a haven for refugees who take up residence in them because of their protected status under international law. In the best cases, refugees help to preserve such sites and defend them from looting, as is now happening in the Ancient Villages of Northern Syria, a World Heritage Site.

• • •

How do we respond to the three *Es*? The three *Ms* frame our humanitarian response for our heritage.

Mitigation. If your basement is prone to flooding, it is probably not the place to store your comic book collection. You can mitigate the risk by storing valuable objects on upper floors. In the same way, caretakers and stakeholders of cultural heritage can plan for and mitigate the impact of many hazards. In the earthquake-prone country of Nepal, staff at the historic Patan Durbar Square retrofitted some buildings with earthquake-resistant construction, an expensive undertaking for which they received international grants. Their efforts were rewarded in 2015 when those buildings withstood the devastating 7.9 Gorhka earthquake with minimal damage while adjacent structures collapsed. Reconstruction of the damaged buildings will cost far more than the investment in earthquake-resistant retrofitting. The caretakers of Patan also had strong local volunteer networks, which quickly organized salvage and safe storage of valuable architectural fragments. In Syria, cultural heritage professionals and activists have managed to evacuate whole museum collections and safeguard immovable heritage in situ. At the Ma'arra Museum in opposition-controlled Syria, staff covered and sandbagged Byzantine-era mosaics to protect them from damage. When the museum was bombed in 2015, the mosaics remained intact.

Movement. In some cases, the only solution to preserving cultural heritage is to move it to a safer place. This involves a great deal of risk. In 2003, staff at the Iraq National Museum secretly emptied their galleries and hid most of their collections ahead of the U.S. military invasion. Looters stole or destroyed thousands of objects when they broke into the museum, but the "secret place" remained intact. During the 2012 Ansar Dine occupation in northern Mali, caretakers there secretly evacuated thousands of the irreplaceable Timbuktu Manuscripts to safety in the capital of Bamako, in the south. And in one of the greatest engineering achievements since the building of the Pyramids, the Egyptian temples of Abu Simbel were disassembled in 1968 and moved to higher ground to save them from being submerged by the Aswan High Dam.

Memory. Documentation is one of the most important responsibilities of cultural heritage caretakers. In the best case, it is fundamental to good collections management practices. In the worst case, when cultural heritage sites are destroyed or looted, documentation is critical to reconstruction or to memorializing what has been lost for scholars and future generations. Advances in 3-D scanning, high-resolution digital photography, and internet databases have made this work easier today than it was in the past. In a dramatic example, archival collections of historic photos and architectural drawings at the Smithsonian's Freer and Sackler Museums document ancient sites in Kathmandu, Aleppo, and Palmyra. All of these images have been digitally scanned in high resolution and made available online, where they are being used by researchers, conservators, and preservation architects.

Cultural heritage is evidence of the best of human activity: artistic genius, historic achievements, and architectural wonders. But at the same time, human activities represent potential hazards that put that heritage at risk. This has been true since ancient times, and over the past hundred years the Anthropocene has introduced new and intensifying threats. Our capacity to destroy cultural heritage sites has reached unprecedented heights, thanks to profound advances in military and industrial technologies, remarkable human population growth, destructive air pollution, and human-induced climate change. To counteract these trends and preserve our cultural heritage we need new research and bold new approaches to disaster risk reduction. Seismic retrofitting and building codes, water mitigation, evacuation of heritage from conflict zones, and more careful military planning: the solutions are often difficult and costly, but the alternative—the loss of identity, stability, and hope for the future—is a much higher price to pay.

THE UNEQUAL ANTHROPOCENE
ROB NIXON

The impossible is the least that one can demand.

—JAMES BALDWIN

"A true ecological approach *always* becomes a social approach," Pope Francis wrote in his 2015 encyclical *Laudato si'*. "[We] must integrate questions of justice in debates on the environment, so as to hear *both the cry of the earth and the cry of the poor*." The pope's words are exhortatory, but they are also indicative of the astonishing rise of environmental justice movements worldwide. The global ascent of activism among the world's most environmentally beleaguered communities is one of the most hopeful stories of our time. Not long ago, those outside the wealthy North commonly dismissed environmentalism as green imperialism, as racist, as antihuman and pro-animal, as an indulgent form of politics that only the comfortable could afford. But in the twenty-first century we are witnessing unprecedented efforts to create coalitions of change, however precarious, between those whom the historian Ramachandra Guha and the economist Juan Martínez-Alier call "full-stomach" and "empty-belly" environmentalists.

The odds of achieving anything resembling justice—for themselves or the environments they depend on—remain stacked against Earth's most impoverished billions. We inhabit an era of short-term, shortsighted plunder, as megacorporations of historically unprecedented wealth, size, and mobility destroy environmental safeguards, creating, in society after society, what George Monbiot calls a globalized "bonfire of regulations." Unanswerable corporations team up with unspeakable autocrats; even in democracies, we witness concerted attacks on public safety nets and the long-term common good. But the fast-moving, deregulated pillage of the most vulnerable ecological and human communities has triggered a pushback, a desperate, determined surge in environmental justice activism, not least in frontline communities at heightened risk from climate collapse. Such communities are already experiencing the Anthropocene's accelerated impacts as an ongoing, staggered trauma.

Empty-belly environmental activism now stretches from the equatorial forests to low-lying Pacific islands, from the Sahel to the Arctic, from the deltas of the Ganges and the Niger to the favelas that have joined the sustainable cities movement. Local environmentalists are creating transnational networks to expose and oppose habitat fracture, biodiversity loss, land seizure, unregulated mining, food and water insecurity, infrastructure deserts, runaway emissions, and the failure to decentralize, decarbonize, and democratize energy access.

Another way of voicing Pope Francis's appeal is to insist that we acknowledge the link between the Great Acceleration and what economists are calling the Great Divide. For the global environmental crisis and the inequality crisis are joined at the hip. In countries as diverse as India, Russia, the United States, Jamaica, South Africa, the United Kingdom, Guatemala, China, and Nigeria, the destabilizing fissure between the megarich and the destitute is widening. In 2010, the 338 most affluent individuals possessed a combined wealth equal to that of the 3.45 billion people who constituted the world's poorest 50 percent. In 2015, a mere sixty-two tycoons matched the combined wealth of the world's poorest half. During those same five years, that club of sixty-two gained $500 billion in personal wealth while the wealth of humanity's poorest half dropped by 41 percent.

In 1962, President John F. Kennedy observed that "our progress in the use of science has been great but our progress in ordering our relations small." His words remain pertinent, not least in the context of Anthropocene debates, where the dominant buzz is about technological innovation, climate engineering, and designing a sustainable future. Far less attention has been paid to the geopolitics of environmentalism's geologic turn. Yes, mitigating the Anthropocene's most destabilizing effects will require technological inventiveness, but the distribution of advances cannot be divorced from questions of political governance and equitable access. Which cabal of engineers gets to decide to reset the global thermostat? Will their experiments be backed by rogue billionaires unanswerable to humanity at large? Will climates be reengineered in the name of humans flourishing and then, in the resource wars to come, be wielded by the rich and for the rich as weapons of mass destruction?

High-consuming humans with energy-intensive lifestyles are leaving knee-deep Anthropocene footprints while billions of others leave so little impress on Earth's life systems that they barely qualify as geomorphic actors. But such deep disparities among human impacts rarely feature in prognoses by Anthropocene techno-optimists, who prefer to operate in the high ether of species thinking. Thus, for the Nature Conservancy's Peter Kareiva and his coauthors, what distinguishes humans from other life-forms is "our unlimited creativity and our sense of moral purpose. The Anthropocene is about designing the future. . . . [It is] an extraordinary opportunity to be welcomed and not feared." The ecologist Erle Ellis cheerily declares that far from being a crisis, the Anthropocene offers a new beginning "ripe with human-directed opportunity." The science journalist Ronald Bailey is sanguine that "over time, we will only get better at being the guardian gods of the earth." The writer Stewart Brand is likewise confident in Anthropocene humanity's surrogate divinity: "We are as gods and might as well get good at it."

Is that splash the sound of Icarus falling into a rapidly warming sea?

Who exactly is the "we" that the Anthropocene's bright-siders love to invoke? It's we the species, big-*H* Humanity, collectively propelled by ethical purpose and technological drive, a superpower whose managerial might is now written in stone. But in a time of deepening divides between concentrated wealth and concentrated abandonment, who will be the unelected deciders, the planetary directors? Which humans will get to stand in for humanity at large? In plutocratic times, the politics of surrogacy can look positively chilling.

The rush toward species thinking that characterizes much Anthropocene thought calls to mind the dispute in the United States that arose around

Black Lives Matter when some outside the movement started proclaiming, "All lives matter." On the surface, this was a generous, inclusive move. But black activists bristled. For the easy universalism of "All lives matter" blurs the focus on systemic discrimination, on the disproportionate burdens of vulnerability borne by black and white Americans. Similarly, viewed through an environmental justice prism, universalizing Anthropocene colloquies about *Homo sapiens* as problem and solution are self-deluding if they fail to address unequal burdens of Anthropocene risk and unequal access to overstrained resources.

Moreover, by positioning humans as bosses of the biosphere, technophiles risk confusing power with control, impact with mastery. Human geomorphic reverberations across Earth's life systems are not synonymous with human dominion over life, which would reduce the infinitely complex interplay among countless animate and inanimate forces to one supreme species' decision making on planetary design. Too often, technophile enthusiasm for the Anthropocene sounds like a hybrid of manifest destiny and selective enlightenment.

To heed, in Pope Francis's words, "the cry of the earth and the cry of the poor," we need to advance alternatives to neoliberalism, which encourages profiteering in the present with little regard for future fallout or social equity. A viable planetary future can be achieved only by attending to the environmental struggles and values of ordinary people, only 16 percent of whom globally live above the U.S. poverty line. The new environmental justice movements, from sustainable cities initiatives to the indigenous Idle No More, are becoming ever more forceful and resourceful. They remain indispensable to any transformative vision of Anthropocene possibility, as they refuse the temporal parochialism, the hubris, and the plutocratic plunder that stand between us and an inclusive, enduring earthly life.

⚹ THE GLOBAL COMMONS

NAOKO ISHII

The organization I lead, the Global Environment Facility (GEF), is arguably the only institution of planetary reach that has embedded in its mission the long-term livability of humanity on a finite, crowded, and increasingly stressed planet. The GEF was created on the eve of the 1992 Rio de Janeiro Earth Summit—itself the cradle of three multilateral environmental agreements, covering climate change, biodiversity, and land degradation—to provide developing countries with the funding necessary to adopt greener development paths in their pursuit of economic growth and poverty eradication.

While some progress in greener development has been made, overall improvement has been woefully insufficient compared to the level of global environment degradation we have witnessed, particularly during the past quarter century. In too many instances, the goals these agreements set were not met or have fallen short of the ambition required to turn the tide in favor of the planet's health. Human activities are rapidly driving the global environment out of the stable conditions humankind has enjoyed for the past ten thousand years.

Stressors that have been accumulating since the Industrial Revolution are now being further magnified by three global megatrends: rapid population growth, a sharp increase of the global middle class, and exploding urbanization. Fewer than four billion people inhabited the world back in 1970; we will number more than nine billion by 2050. By 2030, there will be five billion members of the global middle class. In 1970, about 1.3 billion people lived in urban areas; by 2050, seven billion will be living in cities.

The great majority of the world's scientists have repeatedly warned us of the need for urgent action. In 2014, the Intergovernmental Panel on Climate Change made it clear that the longer we delay tackling climate change, the higher the risks and the costs of dealing with it will be. Others have warned that the planetary system is losing the resilience that has served as a buffer to shocks and other environmental stressors. We may be entering uncharted territory, where abrupt changes—mostly harmful to human societies—cannot be ruled out.

However, there is hope and a way to deal with the unprecedented challenges before us. After a quarter century of feeble attempts, the global community appears to have finally awakened to the call for urgent action. The year 2015 witnessed two landmark accomplishments: the adoption of the Sustainable Development Goals in New York and the climate change agreement at the Conference of the Parties (COP 21) in Paris. Those two historical agreements sent three central messages to the global community.

First, the global environmental commons—be they land, biodiversity, forests, oceans, or the climate system—are the ultimate foundation for sustainable development. Transgressing the planetary boundaries across these dimensions will impose huge costs on humanity.

Second, for humanity to stay within planetary boundaries and hence safeguard the global commons, economies must move to low-carbon development pathways, which will require major changes in our energy, urban, and land-use systems. Incremental progress alone will not suffice.

Third, the solutions needed are everybody's business. It is futile to put all the blame on developed countries while others stay on the sidelines. At the same time, leaving the responsibility solely to governments will also not do the trick. To be crystal clear: without the private sector and the marketplace, transformation will not happen. Perhaps most important, change needs to happen through a global movement in which all citizens can contribute in their own way.

The GEF is ready to take on this challenge and solidify its position as the champion of the global environmental commons. To do that, we have revitalized our strategies and are refocusing our priorities so as to seize the moment and the emerging opportunities before us.

GEF's new strategy, GEF2020, focuses on three principles. The first is a drivers-focused approach. Because our goal is to help transform the systems rapidly being stressed by human activity, the key must be dealing with the root causes of environmental degradation rather than just the symptoms. This means, for example, tackling the drivers of tropical deforestation, which are overwhelmingly associated with agriculture development, in particular a few global commodities such as palm oil, soy, beef, and pulp and paper.

Second, we must adopt an integrated approach to the required interventions. Planetary boundaries do not exist in isolation but instead are highly interdependent. Solutions to the plight of the global commons call for integrated approaches that account for the needs of and interconnections among the many environmental dimensions at the local, regional, and global levels. The multidisciplinary nature of the threats to our climate and to the global environment—and of the solutions needed to counter these threats—is reflected in the mission and the very DNA of the GEF in support of a range of key multilateral environmental accords, including the United Nations Framework Convention on Climate Change, the Convention on Biological Diversity, and the United Nations Convention to Combat Desertification.

Third, we must foster multistakeholder approaches. The root causes of environmental degradation are complex; so are their solutions. Many issues can be effectively tackled only through platforms that can be used to galvanize multiple stakeholders and sectors. For instance, addressing palm-oil-driven deforestation in Indonesia or Malaysia calls for a coalition among all players along the supply chain, including producers large and small, processors, retailers, financial actors, governments, civil society, and consumers at large.

Cities represent another platform for integration. They produce 80 percent of the world's gross domestic product, consume more than two-thirds of the global energy supply, and are responsible for 70 percent of greenhouse gas emissions. In many ways, how we manage our urban areas will largely answer the sustainability question in the Anthropocene world. If managed well, compact, resilient, and resource-efficient cities have the potential to drive the sustainability agenda, contributing to both local livability and global public goods. Interestingly, cities do not have a seat in climate conventions—putting them together is a job for specialized negotiators, representing nation-states. However, cities are where much of the real action on the ground has to happen, and the GEF must stand ready to support them through its engagement with country governments.

COP 21 in Paris captured, to a certain extent, the long-awaited political momentum for climate change. Now is the time to strengthen other political coalitions to address other facets of the global environmental commons, such as the economic value of our oceans and forests, the health of our land and our soils, and the integrity of biodiversity.

A necessary new vision is one that promotes sustainable development by ensuring that our societies thrive within safe planetary boundary limits, thereby avoiding disruptive changes to the climate and other key global systems. Reforms of our major economic systems—food production, energy, and urbanization—are needed that will reflect the reality of operating in the Anthropocene. The GEF, in its role as a financial mechanism for the Rio conventions and a growing number of multilateral environmental agreements spanning virtually all of the planetary boundary dimensions, is uniquely placed to help deal with the complex challenges before us and to buttress Earth's life-support systems in the Age of Humans.

❧ CAN WE REDEFINE THE ANTHROPOCENE?

THOMAS E. LOVEJOY

In 1949, Ruth Patrick, a young limnologist and specialist in the green algae called diatoms, studied the species diversity in the streams and rivers of the mid-Atlantic region of the United States. She concluded that the number and kinds of diatoms (which build beautiful, species-specific silica boxes) reflected not only the natural physics, chemistry, geology, and biology of the watercourses but also the stresses inflicted on their watersheds by human activity. In other words, the biodiversity of a watershed provides an accurate measure of human impact.

That fundamental relationship underpins the water quality programs of the U.S. Environmental Protection Agency to this day. The relationship works because nothing is considered an environmental problem unless it affects living systems, such that, in the end, biodiversity integrates all environmental problems. Sometimes called the Patrick Principle, it applies to terrestrial and marine ecosystems as well as to freshwater systems.

Accordingly, soaring rates of extinction and endangerment are probably the most precise way to measure the impact of humanity on the biosphere. But the Patrick Principle is not just a diagnostic tool; it is equally useful in thinking about how human activities should mesh with landscapes and waterscapes. How might we go from a world that is in trouble biologically, with great global cycles of carbon and nitrogen so distorted that they are impoverishing the diversity of life, to one in which humanity seeks and creates opportunity within the planet's natural fabric?

The interaction between biodiversity and climate change has serious implications for the future of myriad species but, when flipped around, also has the potential to significantly reduce the amount of climate change that life on Earth (including humanity) would otherwise encounter. Often overshadowed by the challenge of fossil fuel emissions is the significant amount of carbon in the atmosphere from centuries of destruction and degradation of modern ecosystems. That carbon can be recaptured through ecosystem restoration. If done at scale, this might even prevent as much as half a degree Celsius of potential temperature increase because of the long lag time between reaching a concentration level of carbon dioxide (CO_2) and the consequent trapping and accumulation of radiant heat.

Restoration in this context primarily means reforestation and restoring grasslands and grazing lands, agricultural ecosystems, and "blue carbon" (coastal wetlands). This will produce additional benefits beyond pulling carbon dioxide out of the atmosphere: for instance, forests provide functioning watersheds, restored grasslands offer better grazing, agroecosystems that accumulate carbon become more fertile, and restored coastal wetlands enhance fisheries and furnish protection against storm surges.

Happily, recognition of the importance of ecosystem restoration is gaining traction. All three of the Rio conventions that sprang from the 1992 United Nations Conference on Environment and Development (the Framework Convention on Climate Change, the Convention to Combat Desertification and Degradation, and the Convention on Biological Diversity) now include restoration as a priority, as does the globally ambitious Bonn Challenge, which emerged from the 2011 meeting of the International Union for Conservation of Nature. And there are multiple other restoration initiatives at various levels of government, civil society, and the private sector.

Implicit in the restoration agenda is the recognition that the planet works not as a physical system alone but as a joint biological and physical system—that it is in fact a living planet and that we need to "manage" it as such. This may sound quite arrogant, as if humans have become carried away with the notion of the Anthropocene, but it really means that we must manage *ourselves*. And it is a way to benefit biodiversity, which is important for myriad reasons. This agenda also represents a profound change in how this human primate could view its place in nature.

So what might restoration look like in a place such as the Amazon—a huge area equivalent to the continental United States, and one of the major repositories of terrestrial biodiversity on the planet? We now know that this vast, unexplored, and seemingly mostly untouched region makes half of its own rainfall. Air masses come off the tropical Atlantic and drop rain, which then evaporates off the complex surfaces of the forest and is transpired through the leaves, returning moisture to the westward-moving air mass, which again falls as rain and repeats the cycle farther to the west.

Prior to this discovery, the prevailing dogma was that vegetation is a consequence of climate and has no influence on it. That belief is now shattered, and echoes of this kind of climate-vegetation interaction have been found in other areas. In the Amazon, it raised a question as soon as the hydrological cycle was recognized: how much deforestation would cause this cycle to degrade and simultaneously reduce the Amazon's capacity to support tropical rain forests and all their amazing biodiversity?

This observation reinforces the idea that the Amazon has to be managed as the system that it is. The actual tipping point of Amazon dieback from failure of the hydrological cycle is not known but is probably in the vicinity of 20 percent deforestation (about the current level). There is no point in discovering the precise tipping point by tipping it, so the sensible approach would be to engage in some reforestation, which the Brazilian government is planning, to build back a margin of safety.

Managing the Amazon as a living system is a complex challenge. Even though more than 50 percent of the region is currently under some form of protection—a remarkable achievement—different sectors (e.g., transportation and energy) reach decisions largely without regard for or oblivious to the implications for further deforestation (through spontaneous colonization, pursuit of short-term economic interests, etc.). The fact that eight nations share the Amazon presents another challenge in itself: even though an Amazon Cooperation Treaty exists, it is largely quiescent and weak.

Although the details of managing the Amazon as a system are complex, they can help us think about managing large units of landscape in general. For that to be successful from a biodiversity perspective, there is a need for much more connection in the natural landscape. Even extensive protected areas, such as Yellowstone National Park in the western United States, need natural connections to the broader landscape. In a changing climate, these connections are even more important because individual plants and animals require relatively natural habitats to move through to track their required climatic conditions. Restoring riparian vegetation along watercourses, for example, goes a long way toward providing the necessary biological connections in the landscape. Protection without connection is insufficient.

Implicit in all of this is a very different model for the interaction of people and nature. The old model, with protected areas set in a matrix of human-dominated landscapes, must be succeeded by one in which human aspiration is embedded in nature. Obviously this model will be harder to apply in more densely populated areas (e.g., many parts of India and southern Asia), but even multilane highways can include overpasses or underpasses for wildlife, and Singapore has been innovative in bringing nature back into the city itself, such as by encouraging natural vegetation at different levels above the ground, not just at street level.

This vision is fundamentally different from its predecessors and much more akin to Edward Hicks's *Peaceable Kingdom*, a series of paintings in which humans and wild animals live in harmony. It will entail abandoning the hubris and environmental destruction (however unaware) that led to this era being named the Anthropocene, and exchanging them for a sense of wonder at and respect for the extraordinary living world of which we are a part. If we do that, the definition of the Anthropocene will change from one in which we are a destructive force to one in which we and succeeding generations on this astounding planet can take true pride in the care we give it.

AFTERWORD

EDWARD O. WILSON

Environmental science is the key to the survival of both the living and the nonliving parts of Earth; it's a complex domain of disciplines still in their infancy. Environmental science also opens the door of science to everyone, and especially to young people, whose commitment to science and technology in service to the environment is vital to our species and the rest of life.

I'm going to use this valuable short space to summarize what scientists have learned about biodiversity and extinction during the past twenty years of fast-moving research. Then I'll suggest what I believe is the only viable solution to halt the continuing and growing rate of species extinction and, thereby, to save the living world.

So, what is biodiversity? What is the living world? It's the collectivity of all inherited variation in any particular given place, such as a vacant lot in a city, an island in the Pacific, or the entire planet. Biodiversity consists of three cascading levels: an ecosystem such as a pond, forest patch, or coral reef; the species composing each ecosystem; and the genes that prescribe the traits that distinguish the species that compose the living part of the ecosystem.

How *much* biodiversity exists? How many species are known to science in the whole world? At the present time, it's almost exactly two million. How many species are there actually still living on Earth, both known and unknown? Excluding bacteria plus the archaea, which I like to call the dark matter of biology because so very little is known of their biodiversity, the best estimate of the number of species of the remaining groups (that is, the fungi, algae, plants, and animals) is about ten million, give or take a million. We know about the big animals—the vertebrates, comprising 63,000 known species collectively of mammals, birds, reptiles, amphibians, and fishes—plus about 270,000 known species of flowering plants. However, next to nothing is known of the millions of kinds of fungi, algae, and invertebrates. These are the foundation of the biosphere. They're the mostly neglected little things that run Earth.

To put the whole matter in a nutshell, we live on a little-known planet. We really don't know what we're doing. At the present rate of elementary exploration, in which about eighteen thousand new species are described and given a Latinized name each year, biologists will complete a census of Earth's biodiversity only sometime in the twenty-third century.

Next, what is the extinction rate around the world? With the data sets of the best-known vertebrate animal species and with additional information from paleontology and genetics, we can put the extinction rate at the closest power of ten: a thousand times greater than the extinction rate that existed before the coming of humans. For example, from 1895 to the present, human activity has driven to extinction fifty-seven freshwater fish species and distinct geographic subspecies in the United States. These extinctions have removed roughly 10 percent of the total diversity previously alive. This means that the extinction rate of fish species in our country is just under an estimated nine hundred times the level that existed before the coming of humans.

This brings me to the effectiveness of the global conservation movement, a great contribution to world culture pioneered by the United States. The global conservation movement has raised public awareness and stimulated a great deal of research. But what has it accomplished in saving species, hence biodiversity? The answer, from data on the vertebrates, is that it has slowed the rate of species extinction but is still nowhere close to stopping it. An expert study of different groups of land vertebrates around the world, species by species, mammals, birds, reptiles, amphibians, found that the rate of extinction in these most favored groups has been cut by about one-fifth,

20 percent. Furthermore, the Endangered Species Act of 1973, by focusing on recognized endangered vertebrates in the United States, with legal procedures and with actions designed for each species in turn, has brought ten times more species back to health than have been lost to extinction.

Nevertheless, the species, and with them the whole of biodiversity, continue to hemorrhage badly. The prospects for the rest of this century are grim. Almost everyone, I'm sure, knows about the 2°C (3.6°F) threshold, the increase in the ground average temperature above which the planet is going to enter a regime of really dangerous climate changes. What is *not* well known is that a parallel situation exists in the living world.

Earth is at or very close to an extinction rate of one thousand times prehuman levels, and the rate is accelerating. Repeat: accelerating. Somewhere between one thousand times and ten thousand times, Earth's natural ecosystems will reach the equivalent of the 2°C global warming threshold and begin to disintegrate, as half or more of the species pass into extinction. We're in the situation of surgeons in an emergency room who've slowed the bleeding of an accident patient by 20 percent. We can say sincerely, "Congratulations! The patient will be dead by morning."

There is a momentous moral decision confronting humanity today. It can be put in the form of a question: what kind of a species, what kind of an *entity*, are we, to treat the rest of life so cheaply? What will future generations think of those now alive having made an irreversible decision of this magnitude so *carelessly*? The five previous such mass extinctions—the last one occurring sixty-five million years ago and ending the Age of Reptiles—required variously five million to forty million years to recover from.

Does any serious person really believe that we can just let the other ten million or so species drain away, and our descendants will somehow be smart enough to take over the planet and ride it like the crew of a real spaceship? That later generations will find the way to equilibrate the land, sea, and air in the biosphere, on which humans absolutely depend, in the absence of most of the natural biosphere?

Most experts understand that only by taking global conservation to a new level can the hemorrhaging of species be brought down to near the original baseline rate, which in prehuman times was one species extinction per one to ten million species per year. Loss of natural habitat is the primary cause of biodiversity extinction—ecosystem, species, and genes, all of it. Only the preservation of much more natural habitat than hitherto envisioned can bring extinction close to a sustainable level. The number of species that can persist in a habitat increases somewhere between the third and fifth roots of the area of the habitat, in most cases close to the fourth root. At the fourth root, a 90 percent loss in area, which is frequent in present-day practice, there will be an automatic drop over time to half the species.

At present, about 15 percent of the global land surface and 3 percent of the global ocean surface are protected in nature reserves. Not only will most of them continue to suffer diminishment of their faunas and floras, but extinction will accelerate overall as the remaining unprotected wildlands and marine habitats shrink because of human activity.

The only way to save the rest of life is to increase the area of protected and inviolable habitat to a safe level. All the signs show that a safe level that can be managed with a stabilized global population of ten billion people is about half Earth's land surface plus half the surface of the sea.

Now, before you start making a list of why it can't be done, why half can't be set aside for the other ten million or so species sharing the planet with us, let me assure you that it most certainly *can* be done, if enough people wish it so.

Think of humanity in this century, if you will, as passing through a bottleneck of overpopulation and environmental destruction. At the other end, if we pass through safely and bring most of the rest of life with us, human existence could be a paradise compared to today. And a long geologic lifespan, essentially immortality, for our species would be possible.

NOTES

Thinking Like a Mountain in the Anthropocene

Page 18, "fierce green fire": Aldo Leopold, *A Sand County Almanac* (New York: Oxford University Press, 1949), 130.

Page 18, "watched the face of many a newly wolfless mountain": Ibid.

Rethinking Economic Growth

Page 32, "the best news in the world today": Jim Yong Kim, "The Lessons of Carabayllo: Making Tough Choices" (plenary address, World Bank Group / IMF Annual Meetings, Lima, Peru, October 9, 2015), www .worldbank.org/en/news/speech/2015/10/09/speech-by-world-bank -group-president-jim-yong-kim-the-lessons-of-carabayllo-making-tough -choices.

The Fire That Made the Future

Page 40, "I sell here": Matthew Boulton (1776), quoted in James Boswell, *The Life of Samuel Johnson, LL.D.*, vol. 2 (London: Henry Baldwin, 1791), 32.

Locating Ourselves in Relation to the Natural World

Page 48, "A certain way of understanding": Jorge Mario Bergoglio [Pope Francis], *Laudato si': Encyclical Letter of the Holy Father Francis on Care for Our Common Home* (Vatican City: Vatican Press, 2015), 75, http:// w2.vatican.va/content/dam/francesco/pdf/encyclicals/documents/papa -francesco_20150524_enciclica-laudato-si_en.pdf.

Page 48, "to become painfully aware": Ibid., 16.

Page 49, "what nature itself allowed": Ibid., 79.

Page 50, "To the Insects": From *The Rain in the Trees* by W. S. Merwin, copyright © 1988 by W. S. Merwin. Used by permission of the Wylie Agency LLC and Alfred A. Knopf, an imprint of the Knopf Doubleday Publishing Group, a division of Penguin Random House LLC. All rights reserved. Any third-party use of this material, outside of this publication, is prohibited. Interested parties must apply directly to Penguin Random House LLC.

Page 51, "The Laughing Thrush": From *The Shadow of Sirius* by W. S. Merwin, copyright © 2008 by W. S. Merwin. Reprinted with the permission of the Wylie Agency LLC and The Permissions Company, Inc., on behalf of Copper Canyon Press, www.coppercanyonpress.org.

Beyond the Biosphere

Page 67, "the extended ecological heliosphere": Valerie A. Olson, "NEO-ecology: The Solar System's Emerging Environmental History and Politics," in *New Natures: Joining Environmental History with Science and Technology Studies*, ed. Dolly Jørgensen, Finn Arne Jørgensen, and Sara B. Pritchard (Pittsburgh: University of Pittsburgh Press, 2013), 197.

Black and Green

Page 84, "Black people are land people": John A. Williams, *The Junior Bachelor Society* (Garden City, NY: Doubleday, 1976), 187.

The City in the Sea

Page 108, "The urban revolution of the Anthropocene": Gaia Vince, "Smart Cities: Sustainable Solutions for Urban Living," BBC Future, May 2, 2013, www.bbc.com/future/story/20130502-how-to-make-our-cities-smarter.

African Art and the Anthropocene

Page 113, "environmental justice thinking": Heather Davis and Etienne Turpin, "Art and Death: Lives between the Fifth Assessment and the

Sixth Extinction," in *Art in the Anthropocene: Encounters among Aesthetics, Politics, Environments, and Epistemologies*, ed. Davis and Turpin (London: Open Humanities, 2015), 7.

Page 113, "Today the talk of the world": Jerry Buhari, artist's statement, *Man and Earth* exhibition, Goethe-Institut, Lagos, Nigeria, November 21–December 3, 2009.

Page 114, "running tears for an irreparable loss": Georgie Papageorge, "Kilimanjaro/Coldfire: Origins of RIFT—Ordered Chaos to a Disordered Present," in *Kilimanjaro/Coldfire: Georgia Papageorge* (Pretoria: Pretoria Art Museum; London: Art First, 2009), 5.

Page 115, "When it comes to speaking": Fabrice Monteiro, quoted in Zahra Jamshed, "'The Prophecy': Photographer Captures Terrifying Vision of Future," CNN, November 17, 2015, http://edition.cnn.com/2015/11/17/arts/photographer-fabrice-monteiro-the-prophecy/.

Page 116, "a 'beyond toxicity' perspective": Thomas J. Doherty and Susan Clayton, "The Psychological Impacts of Global Climate Change," *American Psychologist* 66 (May–June 2011): 272.

Why Polar Bears?
Page 119, "leave the safety of their dens": Marla Cone, *Silent Snow: The Slow Poisoning of the Arctic* (New York: Grove, 2005), 55.

Page 120, "desperate polar bears": Naomi Klein, speaking in *This Changes Everything*, directed by Avi Lewis (Klein Lewis Productions and Louverture Films, 2015), 0:01:00–0:01:11.

Picturing Planetary Peril
Page 129, "I can't think of another movie": A. O. Scott, "Warning of Calamities with a Scholarly Tone," *New York Times*, May 24, 2006.

Page 130, "a nature walk through the book of Revelations": Al Gore, speaking in *An Inconvenient Truth*, directed by Davis Guggenheim (Hollywood: Paramount Pictures, 2006).

Page 132, "I don't know about you": Michael Pollan, "Why Bother?," *Saturday Evening Post*, September–October 2008, 42.

Page 132, "it's not light bulbs": Jeremy Osborn, quoted in Elana Schor, "The Education of a Climate Upstart with a 'Weird' Name," *E&E Daily*, December 13, 2013, www.eenews.net/stories/1059991802.

Dragons in the Greenhouse

Page 138, "the projected impacts of climate change": CNA Military Advisory Board, *National Security and the Accelerating Risks of Climate Change* (Alexandria, VA: CNA Corporation, 2014), 2.

Page 140, "While climate change alone": Ibid., 13.

Why Scientists and Engineers Must Work Together

Page 144, "James Smithson was well aware": Joseph Henry, "On the Smithsonian Institution" (August 1853), in *Proceedings of the Third Session of the American Association for the Advancement of Education* (Newark, NJ: n.p., 1854), 101.

The Unequal Anthropocene

Page 149, "The impossible is the least": James Baldwin, *The Fire Next Time* (New York: Vintage International, 1993), 104.

Page 149, "A true ecological approach": Jorge Mario Bergoglio [Pope Francis], *Laudato si': Encyclical Letter of the Holy Father Francis on Care for Our Common Home* (Vatican City: Vatican Press, 2015), 35, http://w2.vatican.va/content/dam/francesco/pdf/encyclicals/documents/papa-francesco_20150524_enciclica-laudato-si_en.pdf.

Page 149, "full-stomach": Ramachandra Guha and Juan Martínez-Alier, *Varieties of Environmentalism: Essays North and South* (London: Earthscan, 1997), xxi.

Page 149, "bonfire of regulations": George Monbiot, "This Tory Bonfire of Regulations Lets the Rich Foul the Poor with Impunity," *Guardian*,

July 12, 2010, www.theguardian.com/commentisfree/2010/jul/12/tory
-bonfire-regulations-rich-foul-poor.

Page 150, "our progress in the use of science": John F. Kennedy, "Message
from President John F. Kennedy to the *Bulletin of the Atomic Scientists*,"
Bulletin of the Atomic Scientists 18, no. 10 (1962): 2.

Page 151, "our unlimited creativity": Peter Kareiva, Robert Lalasz, and
Michelle Marvier, "Conservation in the Anthropocene," *Breakthrough
Journal* 2 (Fall 2011): 11.

Page 151, "ripe with human-directed opportunity": Erle Ellis, "The Planet
of No Return: Human Resilience on an Artificial Earth," *Breakthrough*
(Winter 2012): http://thebreakthrough.org/index.php/journal/past
-issues/issue-2/the-planet-of-no-return.

Page 151, "over time, we will only get better": Ronald Bailey, "Better to Be
Potent Than Not," *New York Times*, May 23, 2011, www.nytimes.com/
roomfordebate/2011/05/19/the-age-of-anthropocene-should-we-worry/
better-to-be-potent-than-not.

Page 151, "We are as gods": Steward Brand, "The Purpose of *The Whole
Earth Catalog*," *Whole Earth Catalog* 1010 (Fall 1968): www.wholeearth
.com/issue/1010/article/196/the.purpose.of.the.whole.earth.catalog.

FURTHER READING

Alley, Richard B. *Earth: The Operators' Manual*. New York: W. W. Norton, 2011.

Archer, David. *The Long Thaw: How Humans Are Changing the Next 100,000 Years of Earth's Climate*. Princeton, NJ: Princeton University Press, 2009.

Biello, David. *The Unnatural World: The Race to Remake Civilization in Earth's Newest Age*. New York: Scribner, 2016.

Crate, Susan A., and Mark Nuttall, eds. *Anthropology and Climate Change: From Actions to Transformations*. 2nd ed. New York: Routledge, 2016.

Davis, Heather, and Etienne Turpin, eds. *Art in the Anthropocene: Encounters among Aesthetics, Politics, Environments, and Epistemologies*. London: Open Humanities Press, 2015.

Davis, Wade. *The Wayfinders: Why Ancient Wisdom Matters in the Modern World*. Toronto: House of Anansi, 2009.

Del Tredici, Peter. *Wild Urban Plants of the Northeast: A Field Guide*. Ithaca, NY: Cornell University Press, 2010.

Dunaway, Finis. *Seeing Green: The Use and Abuse of American Environmental Images*. Chicago: University of Chicago Press, 2015.

Ellis, Richard. *The Empty Ocean: Plundering the World's Marine Life*. Washington, DC: Island Press, 2003.

Epstein, Paul R., and Dan Ferber. *Changing Planet, Changing Health: How the Climate Crisis Threatens Our Health and What We Can Do about It*. Berkeley: University of California Press, 2011.

Flannery, Tim. *Atmosphere of Hope: The Search for Solutions to the Climate Crisis*. New York: Atlantic Monthly Press, 2015.

Gore, Al. *An Inconvenient Truth: The Planetary Emergency of Global Warming and What We Can Do about It*. Emmaus, PA: Rodale, 2006.

Grinspoon, David Harry. *Earth in Human Hands: The Rise of Terra Sapiens and Hope for Our Planet*. New York: Grand Central, 2016.

Hamilton, Clive. *Earthmasters: The Dawn of the Age of Climate Engineering*. New Haven, CT: Yale University Press, 2013.

Hamilton, Clive, Christophe Bonneuil, and François Gemenne, eds. *The Anthropocene and the Global Crisis: Rethinking Modernity in a New Epoch*. London: Routledge, 2015.

Harvey, David C., and Jim Perry, eds. *The Future of Heritage as Climates Change: Loss, Adaptation, and Creativity*. New York: Routledge, 2015.

Hulme, Mike. *Can Science Fix Climate Change?* Cambridge: Polity, 2014.

Jacobson, Mark Z. *Air Pollution and Global Warming: History, Science, and Solutions*. 2nd ed. Cambridge: Cambridge University Press, 2012.

Jamieson, Dale. *Reason in a Dark Time: Why the Struggle against Climate Change Failed—and What It Means for Our Future*. New York: Oxford University Press, 2014.

Kolbert, Elizabeth. *The Sixth Extinction: An Unnatural History*. New York: Henry Holt, 2014.

Krupnik, Igor, and Dyanna Jolly, eds. *The Earth Is Faster Now: Indigenous Observations of Arctic Environmental Change*. 2nd ed. Fairbanks, AK: Arctic Research Consortium of the United States, 2010.

Leopold, Aldo. *A Sand County Almanac*. New York: Oxford University Press, 1949.

Levy, Barry S., and Jonathan A. Patz, eds. *Climate Change and Public Health*. New York: Oxford University Press, 2015.

Mangelsen, Thomas D. *Polar Dance: Born of the North Wind*. Omaha: Images of Nature, 1997.

Marris, Emma. *Rambunctious Garden: Saving Nature in a Post-Wild World*. New York: Bloomsbury, 2011.

Marsh, Joanna. *Alexis Rockman: A Fable for Tomorrow*. London: D. Giles, 2010.

Martinez-Alier, Joan. *The Environmentalism of the Poor: A Study of Ecological Conflicts and Valuation.* Cheltenham, UK: Edward Elgar, 2002.

Maslin, Mark. *Climate Change: A Very Short Introduction.* 3rd ed. Oxford: Oxford University Press, 2014.

McNeill, J. R. *Something New under the Sun: An Environmental History of the Twentieth-Century World.* New York: W. W. Norton, 2000.

McNeill, J. R., and Peter Engelke. *The Great Acceleration: An Environmental History of the Anthropocene since 1945.* Cambridge, MA: Harvard University Press, 2014.

Milbourne, Karen E. *Earth Matters: Land as Material and Metaphor in the Arts of Africa.* New York: Monacelli, 2014.

Minteer, Ben A., and Stephen J. Pyne, eds. *After Preservation: Saving American Nature in the Age of Humans.* Chicago: University of Chicago Press, 2015.

Möllers, Nina, Christian Schwägerl, and Helmuth Trischler, eds. *Welcome to the Anthropocene: The Earth in Our Hands.* Munich: Deutsches Museum, 2015.

Nakashima, Douglas, et al. *Weathering Uncertainty: Traditional Knowledge for Climate Change Assessment and Adaptation.* Paris: UNESCO, 2012.

Nixon, Rob. *Slow Violence and the Environmentalism of the Poor.* Cambridge, MA: Harvard University Press, 2011.

Nordhaus, William. *The Climate Casino: Risk, Uncertainty, and Economics for a Warming World.* New Haven, CT: Yale University Press, 2013.

Potts, Richard, and Christopher Sloan. *What Does It Mean to Be Human?* Washington, DC: National Geographic, 2010.

Purdy, Jedediah. *After Nature: A Politics for the Anthropocene.* Cambridge, MA: Harvard University Press, 2015.

Pyne, Stephen J. *Between Two Fires: A Fire History of Contemporary America.* Tucson: University of Arizona Press, 2015.

————. *Fire: Nature and Culture.* London: Reaktion Books, 2012.

Roberts, Callum. *The Ocean of Life: The Fate of Man and the Sea.* New York: Viking, 2012.

Sachs, Jeffrey D. *The Age of Sustainable Development*. New York: Columbia University Press, 2015.

Schmidt, Gavin, and Joshua Wolfe. *Climate Change: Picturing the Science*. New York: W. W. Norton, 2009.

Scott, Andrew C., et al. *Fire on Earth: An Introduction*. Chichester, UK: John Wiley and Sons, 2014.

Searles, Harold F. "Unconscious Processes in Relation to the Environmental Crisis." *Psychoanalytic Review* 59 (Fall 1972): 361–74.

Stager, Curt. *Deep Future: The Next 100,000 Years of Life on Earth*. New York: Thomas Dunne Books, 2011.

Steffen, Will, et al. *Global Change and the Earth System: A Planet under Pressure*. Berlin: Springer, 2004.

United Nations Environment Programme. *Measuring Progress: Environmental Goals and Gaps*. Nairobi: United Nations Environment Programme, 2012.

Vince, Gaia. *Adventures in the Anthropocene: A Journey to the Heart of the Planet We Made*. Minneapolis: Milkweed Editions, 2014.

Wagner, Gernot, and Martin L. Weitzman. *Climate Shock: The Economic Consequences of a Hotter Planet*. Princeton, NJ: Princeton University Press, 2015.

White, Gregory. *Climate Change and Migration: Security and Borders in a Warming World*. New York: Oxford University Press, 2011.

Wildcat, Daniel R. *Red Alert! Saving the Planet with Indigenous Knowledge*. Golden, CO: Fulcrum, 2009.

Williams, Mark, Jan Zalasiewicz, Alan Haywood, and Mike Ellis, eds. "The Anthropocene: A New Epoch of Geological Time?" Special issue, *Philosophical Transactions of the Royal Society A: Mathematical, Physical and Engineering Sciences* 369, no. 1938 (March 2011).

Wilson, Edward O. *Half-Earth: Our Planet's Fight for Life*. New York: W. W. Norton, 2016.

CONTRIBUTORS

Richard B. Alley is Evan Pugh University Professor of Geosciences and an associate of the Earth and Environmental Systems Institute at the Pennsylvania State University. He has ranged from Antarctica to Greenland to learn the history of Earth's climate and whether the great ice sheets will melt rapidly, raising sea levels. With more than 290 scientific publications, he has received numerous awards and recognitions, including election to the U.S. National Academy of Sciences and the Royal Society. He hosted the PBS miniseries *Earth: The Operators' Manual.*

Subhankar Banerjee is the Lannan Chair of Land Arts of the American West and a professor of art and ecology in the University of New Mexico's Department of Art and Art History. He is the author of *Arctic National Wildlife Refuge: Seasons of Life and Land* and the editor of *Arctic Voices: Resistance at the Tipping Point.* His photographs have been exhibited in more than fifty museum exhibitions around the world. He is the recipient of several honors, including a Cultural Freedom Award from the Lannan Foundation and a Greenleaf Artist Award from the United Nations Environment Programme.

Carter J. Brandon is the World Bank's global lead economist for environment and natural resources. His primary interest is the linkages among the environment, welfare, and growth. During his twenty years at the World Bank, he has held positions in both headquarters (as the lead economist for Latin America and South Asia) and field offices (in Buenos Aires and Beijing). Prior to that, he ran the Development Economics Group, an economics consulting firm specializing in trade and sector policy analysis. He has a BA from Harvard University and an MSc from Oxford University, where he was a Rhodes Scholar.

Lonnie G. Bunch III is the founding director of the National Museum of African American History and Culture and was previously the president of the Chicago Historical Society and the associate director for curatorial affairs at the National Museum of American History. His publications include *The American Presidency: A Glorious Burden, Call the Lost Dream Back: Essays on Race, Museums and History,* and *Memories of the Enslaved: Voices from the Slave Narratives.*

Paula Caballero is the global director of the World Resources Institute's Climate Program. She has a long history in the field of development, including service as the senior director of the World Bank Group's Environment and Natural Resources Global Practice and as the director for Economic, Social and Environmental Affairs in Colombia's Ministry of Foreign Affairs. She is widely recognized as the lead proponent of the UN's Sustainable Development Goals, for which she was awarded a Zayed International Prize in 2014.

Kelly Chance is a senior physicist at the Smithsonian Astrophysical Observatory and the principal investigator for the NASA/Smithsonian Tropospheric Emissions: Monitoring of Pollution (TEMPO) satellite instrument that is currently being built to measure North American air pollution hourly from geostationary orbit. He has been measuring Earth's atmosphere from balloons, aircraft, the ground, and especially satellites since receiving his PhD from Harvard in 1977. His measurements include the physics and chemistry of the stratospheric ozone layer, climate-altering greenhouse gases, and atmospheric pollution.

Robin L. Chazdon is a professor emerita of ecology at the University of Connecticut. Her long-term collaborative research focuses on successional pathways, vegetation dynamics, and functional ecology of trees in tropical forests. She is currently the executive director of the Association for Tropical Biology and Conservation and the director of the People and Reforestation in the Tropics Research Coordination Network. She works as a consultant on global and regional restoration initiatives and is the author of more than 140 peer-reviewed scientific articles and the coeditor of two books. Her book *Second Growth: The Promise of Tropical Forest Regeneration in an Age of Deforestation* was published in 2014.

Lindsay L. Clarkson is a psychoanalyst in private practice in Chevy Chase, Maryland. A graduate of Harvard University and Duke University Medical School, she is a training and supervising analyst at the Washington Center for Psychoanalysis and a member of the Center for Advanced Psychoanalytic Studies in Princeton, New Jersey. Her primary interests lie in contemporary Kleinian theory and technique and in developing a psychoanalytic understanding of human relationship to the natural world and the non-human environment.

G. Wayne Clough served as the tenth president of the Georgia Institute of Technology from 1994 to 2008 and as the twelfth secretary of the Smithsonian Institution from 2008 to 2014. He earned his BS and MS degrees from Georgia Tech and his PhD from the University of California, Berkeley, and has received honorary doctorates from twelve universities. Clough is active as a lecturer and a thought leader on engineering and climate change. He resides in his native state of Georgia, where he teaches part-time at Georgia Tech, with a focus on leadership, and continues his work to improve higher education access for underresourced students.

Wade Davis is a professor of anthropology and the BC Leadership Chair in Cultures and Ecosystems at Risk at the University of British Columbia. Between 1999 and 2013, he served as an explorer-in-residence at the National Geographic Society. The author of twenty books, including *One River, The*

Wayfinders, and *Into the Silence*, he holds degrees in anthropology and biology and a PhD in ethnobotany, all from Harvard University. His many film credits include *Light at the Edge of the World*, an eight-hour documentary series written and produced for the NGS. Davis, who was made a Member of the Order of Canada in 2016, is the recipient of eleven honorary degrees.

Peter Del Tredici recently retired from the Arnold Arboretum of Harvard University, where he published on the ecology and horticulture of a wide array of trees and shrubs over the course of his thirty-five-year career. He is currently a visiting lecturer in the Department of Urban Studies and Planning at the Massachusetts Institute of Technology, where he teaches courses on urban ecology and climate change. In 2013, the Royal Horticultural Society of England awarded him the Veitch Memorial Gold Medal, "in recognition of services given in the advancement of the science and practice of horticulture."

J. Emmett Duffy is the director of the Smithsonian's Tennenbaum Marine Observatories Network and coordinates the MarineGEO program, a global partnership seeking to understand how and why coastal sea life is changing and how that affects the resilience of marine ecosystems. He is a marine biologist who studies sea grass and coral reef ecosystems worldwide and is active in scientific networks and syntheses linking biodiversity to ecosystem and human well-being. He came to the Smithsonian in 2013 after nineteen years at the College of William and Mary's Virginia Institute of Marine Science.

Finis Dunaway is a professor of history at Trent University in Canada. He is the author of *Natural Visions: The Power of Images in American Environmental Reform* and *Seeing Green: The Use and Abuse of American Environmental Images*, which received the John G. Cawelti Award from the Popular Culture Association/American Culture Association, the History Division Book Award from the Association for Education in Journalism and Mass Communication, and the Robert K. Martin Book Prize from the Canadian Association for American Studies. He is currently researching the history of the Arctic National Wildlife Refuge, a remote area in northeastern Alaska that has become one of the most contested landscapes in modern North America.

John Grabowska's natural history films about the American wilderness have won awards at festivals around the world and are broadcast nationally as prime-time specials on PBS. He has lectured on natural history filmmaking at the Smithsonian Institution and the National Geographic Society, led environmental media workshops in Argentina and Panama, and cofounded the American Conservation Film Festival.

Naoko Ishii has served as the CEO and chair of the Global Environment Facility since August 2012. Prior to that, she was Japan's deputy vice minister of finance, responsible for the country's international financial and development policies and its global policies on the environment, including climate change and biodiversity. At the World Bank from 2006 to 2010, she was the country director for Sri Lanka and the Maldives. She has published several books, one of which was awarded the Suntory Prize and one the Okita Memorial Prize for International Development Research. She holds a BA and a PhD from the University of Tokyo.

Luc Jacquet's first feature film, *March of the Penguins*, won the Academy Award for Best Documentary Feature in 2006. In 2010, he founded the nonprofit association Wild-Touch, which aims to support and increase wildlife conservation by arousing emotions through powerful images. In 2014, he directed *Ice and the Sky*, a documentary about the glaciologist Claude Lorius, who contributed to revealing the impact of human activity on climate. In 2017, the director is releasing *The Emperor*, a sequel to his first movie, filmed on his recent "Antarctica!" Wild-Touch expedition. Jacquet and Wild-Touch are also working on "The Flow of Life," an artistic and educational project about biodiversity, which will explore different ecosystems and explain how all living beings on our planet are interconnected.

Elizabeth Kolbert is a staff writer at the *New Yorker*. "The Climate of Man," her three-part series on global warming, won the American Association for the Advancement of Science Journalism Award, the National Academies Communication Award, and the National Magazine Award for Public Interest. She is the recipient of a Lannan Literary Fellowship, a Heinz Award, and a National Magazine Award for Reviews and Criticism. Her books include *The Prophet of Love: And Other Tales of Power and Deceit*, *Field Notes from a Catastrophe: Man, Nature, and Climate Change*, and *The Sixth Extinction: An Unnatural History*, the winner of the 2015 Pulitzer Prize for General Nonfiction.

W. John Kress is a distinguished scientist and curator of botany at the National Museum of Natural History. He formerly served as the interim undersecretary for science and the Grand Challenges Consortia's director of science at the Smithsonian. He received a BA from Harvard University and a PhD from Duke University. His books include *Plant Conservation: A Natural History Approach*, *The Weeping Goldsmith*, *The Art of Plant Evolution*, and *The Ornaments of Life*. He is a fellow of the American Association for the Advancement of Science and an adjunct professor of biology at George Mason University in Virginia.

Igor Krupnik is a curator of the Arctic and Northern Ethnology collections and the head of the Ethnology Division at the National Museum of Natural History. Trained as a cultural anthropologist and ecologist, he has worked for more than four decades in indigenous communities in Alaska and the Bering Strait region. His areas of expertise include modern cultures, indigenous ecological knowledge, and climate change and its impact on the people of the Arctic. He has written and coedited more than twenty books, catalogs, and heritage source books and was the lead science curator for the 2006 Smithsonian exhibition *Arctic: A Friend Acting Strangely*.

Thomas E. Lovejoy became fascinated with biological diversity and the natural world at the age of fourteen and now works on the interface of science and public policy. He first went to the Amazon in 1965 and did his graduate work in those rain forests, earning a PhD from Yale University in 1971. In 1980, he was the first both to use the term *biological diversity* and to make projections of species extinctions. His research program on forest fragments in the Brazilian Amazon is ongoing after thirty-seven years. He has coedited three books on biodiversity and climate change (*Global Warming and Biological Diversity*, *Climate Change and Biodiversity*, and a forthcoming work from Yale University Press).

George E. Luber is the chief of the Climate and Health Program at the National Center for Environmental Health, Centers for Disease Control and Prevention. Since joining the CDC in 2002, he has served as an Epidemic Intelligence Service officer and epidemiologist. He is a convening lead author for the U.S. National Climate Assessment and a lead author for the Intergovernmental Panel on Climate Change's Fifth Assessment Report.

Joanna Marsh is the senior curator of contemporary interpretation at the Smithsonian American Art Museum. From 2007 to 2015, she served as the James Dicke Curator of Contemporary Art at the museum. Prior to joining the Smithsonian, Marsh held curatorial positions at the Wadsworth Atheneum Museum of Art in Hartford, Connecticut. Her most recent exhibitions, *The Singing and the Silence: Birds in Contemporary Art* (2014) and *Alexis Rockman: A Fable for Tomorrow* (2010), focus on the interrelation of art, science, and humanity. She holds a BA from Cornell University and an MA in postwar and contemporary art from Sotheby's Institute, London.

Douglas J. McCauley began his career as a fisherman in the Port of Los Angeles but ultimately migrated to marine science and now serves as an assistant professor at the University of California, Santa Barbara. He has degrees in political science and biology from the University of California, Berkeley, and a PhD from Stanford University. He did postdoctoral research at Stanford, Princeton University, and UC Berkeley. He is an Alfred P. Sloan Research Fellow in the Ocean Sciences and the director of the Benioff Ocean Initiative.

Sean M. McMahon is a senior scientist at the Smithsonian Environmental Research Center and the Temperate Program coordinator for the Smithsonian's Forest Global Earth Observatory. He received an MS in statistics and a PhD in ecology and evolutionary biology at the University of Tennessee, Knoxville. He held a postdoctoral research position at Duke University before joining the Smithsonian in 2010. His research focuses on current and future dynamics of temperate and tropical forests.

J. R. McNeill is a university professor and professor of history at Georgetown University. He has won two Fulbright awards, a Guggenheim fellowship, a MacArthur grant, and a fellowship at the Woodrow Wilson Center. His books include *Something New under the Sun: An Environmental History of the Twentieth-Century World* and *Mosquito Empires: Ecology and War in the Greater Caribbean, 1620–1914*, which won the American Historical Association's Beveridge Prize. In 2010, he was awarded the Toynbee Prize for "academic and public contributions to humanity." He has served as the vice president of the American Historical Association and the president of the American Society for Environmental History.

Karen E. Milbourne has been a curator at the Smithsonian's National Museum of African Art since 2008. Previously, she was an associate curator of African art and the department head for the Arts of Africa, the Americas, Asia and the Pacific Islands at the Baltimore Museum of Art and, prior to that, an assistant professor of art history at the University of Kentucky. Her expertise includes the arts and pageantry of western Zambia and contemporary African art. Milbourne received her PhD in art history from the University of Iowa in 2003 and has been the recipient of numerous awards, including a Fulbright Fellowship.

Rob Nixon is the Currie C. and Thomas A. Barron Family Professor in Humanities and the Environment at Princeton University. From 1999 to 2015, he held the Rachel Carson Professorship in English at the University of Wisconsin, Madison. His writing and teaching have a strong focus on struggles for environmental justice in the Global South. His most recent book, *Slow Violence and the Environmentalism of the Poor*, received four prizes, including an American Book Award.

Ari Novy is the executive director of the U.S. Botanic Garden in Washington, DC, and a research collaborator in the National Museum of Natural History's Department of Botany. He is trained as an evolutionary ecologist and has research experience in a variety of plant-related areas, including invasion biology, conservation, horticultural improvement, beekeeping management, plant evolution, and agricultural economics and policy.

Rick Potts is a paleoanthropologist and the director of the Human Origins Program, based at the National Museum of Natural History. Leading field excavations in the East African Rift Valley and in China, he has researched the effect of environmental instability on human evolution, attracting wide attention and stimulating new studies in earth science, paleontology, and experimental and computational biology. He is a curator of the Smithsonian's Hall of Human Origins and a coauthor of its companion book *What Does It Mean to Be Human?*

Stephen J. Pyne's interest in fire began with fifteen seasons as a North Rim Longshot at Grand Canyon National Park, followed by three seasons writing fire plans for the National Park Service. His *Cycle of Fire* suite is a survey of fire on Earth. His other major books include *The Ice: A Journey to Antarctica*, *How the Canyon Became Grand*, and *Voyager: Exploration, Space, and the Third Great Age of Discovery*. He is a professor in the School of Life Sciences at Arizona State University and a past president of the American Society for Environmental History.

Lisa Ruth Rand earned her doctorate in the history and sociology of science at the University of Pennsylvania in 2016. She is currently a Mellon Postdoctoral Fellow at the University of Wisconsin, Madison. Her research examines the environmental history of near-Earth space, focusing on how space junk became a subject of international environmental concern during the Cold War.

Peter H. Raven served as the director of the Missouri Botanical Garden and the George Engelmann Professor of Botany at Washington University for thirty-nine years (1971–2010). A well-known student of plant systematics and evolution and a leading conservationist, he is a member of many academies of science (United States, United Kingdom, India, China, Russia, and Brazil among them) and the recipient of the U.S. National Medal of Science. He is the coauthor of the globally best-selling botany text *Biology of Plants* and the author and coauthor of many other books and papers.

Torben C. Rick is a curator of human environmental interactions and the chair of the Department of Anthropology at the National Museum of Natural History. His research focuses on the archaeology and historical ecology of coastal and island peoples, especially on the North American Pacific and Atlantic coasts. He has field projects on California's Channel Islands and on the Chesapeake Bay, where researchers from a variety of disciplines (including anthropology, biology, and ecology) collaborate and focus on ancient and modern human environmental interactions.

Holly H. Shimizu was the executive director of the U.S. Botanic Garden in Washington, DC, for more than fifteen years. During that time, in addition to developing major exhibits and programs, she led conservation efforts, including the cofounding of the Sustainable Sites Initiative (SITES). Shimizu has received numerous awards for her work and is currently a teacher, writer, horticultural consultant, and leader of botanical tours.

Jeffrey K. Stine is a curator for environmental history at the Smithsonian's National Museum of American History. He has served as a book series editor at Resources for the Future Press and the University of Akron Press and is a past president of the American Society for Environmental History. His books include *Mixing the Waters: Environment, Politics, and the Building of the Tennessee-Tombigbee Waterway*, *Twenty Years of Science in the Public Interest*, and *America's Forested Wetlands: From Wasteland to Valued Resource*.

Corine Wegener is a cultural heritage preservation officer in the Office of the Provost/Under Secretary for Museums and Research at the Smithsonian Institution and manages the Smithsonian Cultural Rescue Initiative. Since 2010, the SCRI has provided disaster recovery training in Haiti, Syria, Iraq, Egypt, Mali, Nepal, and the United States, as well as cultural heritage training for U.S. military and law enforcement personnel.

Edward O. Wilson is a university research professor emeritus at Harvard University. As one of the world's preeminent biologists and naturalists, he has received more than one hundred national and international awards, including the National Medal of Science and two Pulitzer Prizes for General Nonfiction (for *On Human Nature* and *The Ants*). His most recent book, *Half-Earth: Our Planet's Fight for Life*, concludes his trilogy begun by *The Social Conquest of Earth* and *The Meaning of Human Existence*.

Scott L. Wing was born in New Orleans and grew up there and in Durham, North Carolina. His childhood interest in fossils was reinforced by field-work in Wyoming while he was a student at Yale University. After completing his doctorate there, he was a National Research Council postdoctoral fellow at the U.S. Geological Survey. He moved to the Smithsonian Institution in 1984 and has spent much of his career studying past periods of global warming.

ACKNOWLEDGMENTS

Inspiration for this edited volume sprang from a series of workshops, symposia, and informal discussions organized by scholars and scientists from across the Smithsonian Institution. It has been a pleasure and a remarkable learning experience to work with the dedicated group of colleagues who led the Smithsonian's Living in the Anthropocene initiative, and we are deeply grateful to the entire team, which included, besides us, Bill Allman, Kelly Chance, Pierre Comizzoli, Michelle Delaney, Christine France, Eric Hollinger, Tim Johnson, Christine Jones, Liz Kirby, Robert Leopold, Odile Madden, Sean McMahon, Jennifer McMillan, Karen Milbourne, Jane Passman, Rick Potts, Barbara Rehm, Torben Rick, Corine Wegener, and Scott Wing.

Many others within the Smithsonian contributed to the initiative, among them Martin Collins, Catherine Denial, Bill DiMichele, Bert Drake, Emmett Duffy, Jessica Faison, William Fitzhugh, Johnny Gibbons, Elizabeth Kennedy Gische, Brian Gratwicke, Peter Haydock, Kristofer Helgen, the late Len Hirsch, Peter Jakab, Andrew Johnston, Justin Kasper, Jonathan Kavalier, Igor Krupnik, Erin Kuprewicz, Joanna Marsh, Patrick Megonigal, Whitman Miller, Steven Monfort, Suzan Murray, Eva Pell, Ira Rubinoff, Melissa Songer, Maggie Stone, Gabrielle Tayac, Kristina Anderson Teixeira, Jonathan Thompson, Jim Wood, and Joe Wright. Contributors from outside the Smithsonian included Thad Allen, Richard Alley, Subhankar Banerjee, Lindsay Clarkson, Elizabeth Cottrell, James Fleming, Thomas Friedman, James Hack, Drew Jones, Chris Jordan, Rachel Kyte, Kimberlyn Leary, Thomas Lovejoy, George Luber, Charles Mann, Humphrey

Morris, Donald Moss, Rob Nixon, Sabine O'Hara, Andrew Revkin, Gavin Schmidt, Mary Evelyn Tucker, Woody Turner, Daniel Wildcat, Timothy Wirth, and Lynne Zeavin.

For their ongoing support, we thank Smithsonian Secretary David Skorton, former Smithsonian Secretary Wayne Clough, Smithsonian Acting Provost Richard Kurin, and the directors of our respective museums: Kirk Johnson of the National Museum of Natural History and John Gray of the National Museum of American History. The Living in the Anthropocene initiative benefited from funds provided by the Bill and Melinda Gates Foundation.

In addition, W. John Kress extends his personal thanks to Elizabeth and Phil Ryan for their support of our efforts to understand the Anthropocene and to his wife, Lindsay Clarkson, for her encouragement and engaging discussions. Jeffrey K. Stine is indebted to Jonathan Cobb, Evelyn Hankins, Ezra Heitowit, Lexi Lord, Mark Madison, Harry Rand, Jim Roan, Marc Rothenberg, Michael Brian Schiffer, Roger Sherman, Barbara Stauffer, Deborah Jean Warner, Helena Wright, and—especially—his wife, Marcel Chotkowski LaFollette.

INDEX

beliefs about natural world, 46
Belize, 92
Bellow, Saul, 45
belonging, sense of, 52
benefit-to-cost ratio of reducing pollution, 34
Benin, 115
"beyond toxicity" perspective, 116
bioaccumulation, 119
biodiversity, 32, 90, 93, 157–59, 161
Black Lives Matter, 152
Blue Ridge Mountains, 125
"bonfire of regulations," 149
Bonn Challenge, 158
Boston, 59
bottleneck, environmental, 164
Boulton, Matthew, 40
Brand, Stewart, 151
Brazil, 89, 159
Breakthrough Energy Coalition, 144
British Antarctic Survey, 64
British Broadcasting Corporation, 126
British Columbia, 46
Bronx Zoo (Rockman), 110–12
Brooklyn, 109–11; Brooklyn Bridge, 110
Buhari, Jerry, 113

Cairo, 146
California, 23, 53
Calling All Polar Bears (Warden), 120
calls to action, 77, 81, 116
Canada, 117–18
canyons, urban, 100
carbon dioxide, *12*, 19–20, 38, 62, 101–2, 138, 158
Carbon-Neutral Energy Solutions Laboratory, 143
carbon-neutral energy sources, 94, 144
carbon sequestration, 91, 157–58
cardiovascular health, 102
catastrophic events, 106, 129
Channel Islands (CA), 76
Chapman chemical mechanism, 64
charcoal, 55
Chasing Ice (documentary), 130
chestnut, American, 53

Chicago, 84
China, 34, 139
chlorofluorocarbons, 62, 65
Churchill (Canada), 117–18
cities, 58–61; at-risk areas in, 83; expansion of, 108, 146; heat islands in, 60, 100; horticulture in, 96–98; management of, 155; nonnative species in, 39; population of, *15*, 16; reintroducing wildlife in, 111; sustainability of, 112; topographic settings of, 101; urbanization, 102, 153
citizen-scientists, 93
civilization, 29; traces of, 121, 123, 127
civil rights, 86
Clayton, Susan, 116
Clean Air Act, 34
Clean Energy Investment Center, 144
clear-cutting, 55
climate change: adaptation plans for, 142; agriculture impacted by, 96; Arctic impacted by, 117; and biodiversity, 157; delay in addressing, 153; forests impacted by, 89–90; gradual nature of, 129; health impacts of, 99; indigenous peoples and, 78–82; in natural history films, 127; oceans affected by, 26; psychological dimensions of, 38; systems impacted by, 144; war and, 140. *See also* global warming; greenhouse effect
climate fiction (cli-fi), 109
climate-vegetation interaction, 159
coal, 63
coastlines, 91, 101, 145
co-benefits approach, 102
collaboration, 136
collages, 114
colonization, 44, 122
Columbian Exchange, 15
Columbus, Christopher, 138
Committee on New Orleans Regional Hurricane Protection Projects, 142
commons, global, 153–56
Cone, Marla, 119
conquest of nature, 111, 121–22
consciousness, environmental, 93

Enlightenment, 45
"Enough with the Doom and Gloom" (Lubchenco), 128
Environmental Design Gold, 85
Environmental Protection Agency, U.S., 64, 111–12, 157
Environment and Object: Recent African Art (exhibition), 114
Eocene epoch, 8, 19
ethnocentrism, 44
Eucalyptus regnans, 53
evasion, as a reaction, 48
Evelyn, John, 63
evolutionary time, 2
exotic species. *See* nonnative species
exploration, 140–41
extinction, 3, 20, 24, 32, 76, 157, 162–64
Extreme Ice Survey, 130
extreme weather events, 99, 100, 108
Exxon Valdez, 128

faith, absolute, 45
Fall and Spill History (Buhari), 113
feedback mechanisms, 65
fertility, human, 16
fertilizer consumption, *13*
filled land, 59
films, 106, 109, 121, 124, 126–29
finite supplies of resources, 16
fire, 40–43
fishing, 24, 26, 91–92; capture of marine fish, *11*
Flaherty, Robert, 126
Flint (MI), 83
floods, 84, 101
fog, 63, 140
folk stories, 53
food: cost of, 139; production of, 96–98.
    See also agriculture
Forest Global Earth Observatory (ForestGEO), 56
forests: deforestation, 75, 88–89; homogenization of, 89; livelihoods based on, 90; logging, 46; management of, 55; regeneration of, 55, 87; research on, 56;

secondary, 90, 125; succession in, 73, 87–90; temperate, 53–57; tropical, 57, 88, 90; understanding of, 88
fossil fuels, ix, 38, 41, 54
foxes, 76
fragmentation of land, 59
Francis Little House, 146
free-radical nitrogen and halogen, 64
Freer Museum, 148
Freons, 62, 65
Frozen Planet (Discovery), 127
Fukushima nuclear accident, 25
"full-stomach" environmentalists, 149
Fumifugium (Evelyn), 63
future: artistic representations of, 110–12, 116; desired conditions in, 76; fear of, 122; frank examination of, 4; of humanity, 27; of societies, 30

Galerie Börgmann, 115–16
gardening, 95–98
gas drilling, in Arctic, 120
GDP. *See* gross domestic product
GEF2020 strategy, 154
genes, 161
genetic modification of crops, 95
geologic time, 2, 19, 22
geology, 20
Georgia Institute of Technology, 143
geospatial technology, 92
geothermal power, 102
Giza pyramids, 146
glaciers, 114, 130
Global Change and the Earth System (International Geosphere-Biosphere Programme), 130–31
global commons, 153–56
Global Environment Facility, 153–56
Global Fishing Watch (SkyTruth), 92
Global Footprint Network, 96
global warming, 39, 140. *See also* climate change; greenhouse effect
goals, agricultural, 96
Goethe-Institut, 113
Gore, Al, 118, 128

Gorman, Alice, 66
Gowanus (Rockman), 110–12
grassroots initiatives, 79, 82
Great Acceleration, 8, 10, *11–15*, 16–18, 67–68, 130–31, 150
Great Divide, 150
The Great Transformation (Polanyi), 10
greenhouse effect, 38, 62–63, 99. *See also* climate change; global warming
gross domestic product (GDP), 32
Guggenheim, Davis, 118
Guha, Ramachandra, 149

habitats: clearing of, for agriculture, 95; degradation of, 38; disturbed, 89; features of individual, 79; marine, 73; preservation of, 164
Half-Earth Project, 128
Hampton Institute, 84
Hann Bay (Senegal), 115
Hawaii, 89
hazards to cultural heritage, 145–48
health, human, 73, 99–103
heat, absorption of by oceans, 91
heat islands, urban, 60, 100
heat-related illnesses, 100
heat waves, 99, 101
heliosphere, extended ecological, 67
Henry, Joseph, 144
Henry IV (Shakespeare), 54
Heraclitus, 40
heritage, cultural, 145–48
Hesiod, 43
Hicks, Edward, 160
highways, 84
Holocene epoch, ix, 1, 8, 99
homogenization of forests, 89
Homo sapiens. See humans
hope, in Anthropocene, 127. *See also* optimism, in Anthropocene
Hopkins, Sam, 115–16
horizontal networking, 82
horticulture, 58, 95–98
Howard University, 83
Hudson Bay, 117–18

human rights, 80–86
humans: changes caused by, 1; domination of world by, ix; environments inhabited by, 121; evolutionary success of, 9; extinction of, 21; and forest succession, 88; impact of activities, 8; origins of, 27; resources supporting, 21; teleological narrative of, 131
hunter-gatherers, 75
Hunt-Lenox Globe, 138–41
hurricanes: Katrina, 84, 142; Sandy, 111, 139, 142
Huxwhukw people, 46
hydrological cycle, 159

Idle No More movement, 152
Iliad (Homer), 64
illnesses, human, 100–02
impervious surfaces, 59–60, 100
impoverishment, emotional, 48
An Inconvenient Truth (Gore), 118, 129–30
indigenous peoples, 3, 46–47, 72, 78–82, 106; observations by, 79
individual actions, in Anthropocene, 131
individualism, 45, 132
Indonesia, 155
industrial pollutants, 119
Industrial Revolution, 1, 9, 15, 44
industry, ocean, 25
inequality, 3, 131, 136, 149–52
infrastructure, 101, 102
initiation rites, 46
injustice, 3, 26, 30, 80–81, 131
innovation, 27–28, 144, 150–51
insects, 50–51
integrated approaches, 155
Intergovernmental Panel on Climate Change, 153
internal combustion, 42
internal conflicts, 49
International Geosphere-Biosphere Programme, 16, 130–31
International Union for Conservation of Nature, 24, 158
interpretations by indigenous peoples, 79

Inuit people, 117–20
inundation of coastal areas, 101
invasive species, 61, 122, 125. *See also* nonnative species
invisibility of natural threats, 106
Iraq, 146; Iraq National Museum, 147
iron industry, 54
ISIS. *See* Islamic State of Iraq and the Levant
Islamic State of Iraq and the Levant (ISIL), 146
island fox (*Urocyon littoralis*), 76

Janzen, Dan, 98
Jolly, Dyanna, 117
justice, 80–81, 113, 149–52

Kaktovik (AK), 118
Kareiva, Peter, 151
Kathmandu (Nepal), 148
Kennedy, John F., 150
Kenya, 115–16
keystone predators, 18
Kim, Jim Yong, 32
Klein, Naomi, 120
knowledge, 137–41, 144
Krupnik, Igor, 117
Kwakwaka'wakw people, 46

Lagos (Nigeria), 113
land: fragmentation of, 59; landfill, 59; modifications to, 95; subjected to human will, 123; understanding of, 47; use, 84, 89
landscapes, 58, 88, 97
Langley, Samuel Pierpont, 62
Latour, Bruno, 69
Laudato si' (Pope Francis), 48, 149
"The Laughing Thrush" (Merwin), 51–52
Leadership in Energy certification, 85
Leopold, Aldo, 18
lithic landscapes, 41
livelihoods, forest-based, 90
Living in the Anthropocene project, 144
local scale, 79
Loess Plateau, 34
logging, 46

London, 63
long-term perspectives, 20, 22, 84, 124, 130
Lorentz, Pare, 126
Los Angeles, 63, 83
Lubchenco, Jane, 128

Ma'arra Museum, 147
Maasai Steppe Ascending—Convective Displacement (Papageorge), 114
Malaysia, 155
Mali, 115, 139–40, 146–48
malnourishment, 95
managing the environment, 55–57, 158
Man and Earth (Buhari), 113
Mangelsen, Thomas, 117
Manhattan Bridge, 110
Manifest Destiny (Rockman), 109
marine fish capture, *11*
Marine Global Earth Observatory (Marine GEO), 93
marine habitats, 73
Marine Industrial Revolution, 25
markets, prevalence of, 10
Martínez-Alier, Joan, 149
mass extinctions, 3, 76, 163
materialism, 45
Maya people, 74–75
Mbeubeuss marshes (Dakar), 115
meaning, creation of, 28
mechanism, reduction of world to, 45
megacities, 108
megacorporations, 149
megafauna, 23
memory, distortion of, 51
Menominee nation, 56
mercury, 25
Merwin, W. S., 50–52
metaphors, 45–46
methane, *12*, 62, 139
middle class, 153
migration, 39, 89, 147
minority rights, 80
Mission Innovation agreement, 144
Mississippi, 84
mistrust, 29

Palmyra, 146, 148
pan-African photography biennial, 114–15
Papageorge, Georgia, 114
paper manufacturing, 55
Paris Agreement, 33, 142, 154–56
Partial Test Ban Treaty, 67
particulate matter, 64
Patan Durbar Square, 147
Patrick, Ruth, 157
Patrick Principle, 157
pavement, 59
Peaceable Kingdom (Hicks), 160
persistence of anthropogenic changes, 3
PETM. *See* Paleocene-Eocene Thermal
   Maximum
photochemical smog, 63
photography, 117–18, 130
photovoltaic technology, 143
piñon pine, 127
pirate fishing, 92
planetary perspective, 80, 131
plant-based diets, 102
plants: archaeological study of remains,
   74; cultural significance of, 60; riparian
   vegetation, 160; urban, 60. *See also*
   agriculture; gardening; horticulture
plastics, 25, 92, 115–16
Plato, 43
Pliny the Elder, 40
The Plow That Broke the Plains (Lorentz),
   126
poison ivy (*Toxicodendron radicans*), 101
Polanyi, Karl, 10
polar bears, 106, 117–20
Polar Biology, 118
Polar Dance (Mangelsen), 117
Pollan, Michael, 132
pollen, 101
pollution: air, 99; benefit-to-cost ratio
   of reducing, 34; deaths caused by, 32;
   industrial, 119; in outer space, 67;
   oxidizing, 63; "reducing-type," 63
polychlorinated biphenyls, 119
Pootoogook, Annie, 119
Pope Francis, 48, 149–50, 152

population: density, 101; growth, 10, 88,
   95–96, 153; urban, 16
Population Reference Bureau, 95–96
postapocalyptic visions, 109
Potomac River, 125, 128
poverty, 32, 101
preadaptation, 60
preservation of habitat, 164
private sector, 35
proactive responses to Anthropocene, 81
production, sustainable levels of, 34
Prometheus Bound (Aeschylus), 40
Prometheus myth, 40–43
Prometheus Unbound (Shelley), 41
The Prophecy (Monteiro), 115
protected areas, 159–60, 164
protest films, 127
Providence (RI), 84
psychoanalysis, 48
psychological dimensions of planetary
   change, 38
Public Broadcasting Service, 126
Puerto Rico, 89
pyric transition, 42

Queensland (Australia), 89

race and racism, 72, 86
ragweed (*Ambrosia artemisiifolia*), 101
rapid change, surviving, 3, 21
real-time information, 93
recycling, 84
"reducing-type" pollution, 63
redwoods, 53
Reef Life Survey, 93
refugee crises, 146–47
regeneration of forests, 55, 87
regulations, 149
reintroduction of wildlife, 111
Renaissance, 45
Rencontres (Papageorge), 114
repeat photography, 130
research forests, 56
resilience, 85, 88, 110, 140, 153–54
resistance, indigenous, 120

symbols, use of, 27
Syria, 139, 146–47
systemic change, 132–33

Tanzania, 114
targets for conservation and restoration, 76
technology, 27–28, 35, 91–92, 123, 143, 150–51
teleological narrative of *Homo sapiens*, 131
television, 126
"Thinking Like a Mountain" (Leopold), 18–22
This Changes Everything (Klein), 120
thresholds for rapid change, 140
timber, 54
Timbuktu Manuscripts, 146–48
time, deep sense of, 52
time-lapse photography, 130
Time magazine, 118
tipping points, 159
toleration, role of civilization, 29, 48
Tomb of Jonah (Nebi Junis), 146
tool making, 27
topographic settings, 101
"To the Insects" (Merwin), 50
transitions in Earth history, 3, 8
trauma, images of, 116
trends, accelerating, 11–15
tropical regions, 26, 57, 88, 90
True-Life Adventures (Disney), 126
trust and social web, 29
Turpin, Etienne, 113
Tuskegee Institute, 84

ultraviolet radiation, 64
uncertainty, environmental, 29, 52, 87–90, 138–41
United Nations Conference on Environment and Development, 158
United Nations Convention to Combat Desertification, 155
United Nations Earth Summit, 80
United Nations Educational, Scientific and Cultural Organization (UNESCO), 114, 145–46

United Nations Framework Convention on Climate Change, 155
United States, 66–67, 72, 89, 110, 151–52, 162
universalism, 131, 152
University Center of Excellence in Photovoltaic Research and Education, 143
urban areas. *See* cities

values, shared, 30
Venice, 145
Vermont, 113
videos, 106, 114
Vince, Gaia, 108
visual culture, 3, 107, 129–33
volcanoes, 146
Voltaire, 98
volunteers and research, 93
vulnerabilities of natural systems, 78, 100, 131, 139

war, 139–40, 146
Warden, Allison Akootchook, 120
Washington, DC, 83, 142
water: increase in use of, *13;* regulation of, 90
watersheds, 157
wealth inequality, 32, 150
westward expansion, 110
wilderness, 23, 54, 91
wildflowers, 128
wildlife, reintroduction of, 111
Williams, John, 84
Wilson, E. O., 128, 161–64
wind power, 102
Wisconsin, 56
wonder, sense of, 126
World Bank Group, 32
World Economic Forum, 32
World Health Organization, 32
Wright, Frank Lloyd, 146

Yellowstone National Park, 159–60
yields, agricultural, 95

zoos, 111